Managing Your Career Success

Managing
Your
Career Success

Practical Strategies for Engineers, Scientists, and Technical Managers

TERRY D. SCHMIDT

Lifetime Learning Publications

A division of Wadsworth, Inc.
Belmont, California

London, Singapore, Sydney, Toronto, Mexico City

Jacket and Text Designer: Richard Kharibian
Copy Editor: Susan McCalla
Illustrator: John Foster
Composition: Graphic Typesetting Service

Printed in the United States of America

1 2 3 4 5 6 7 8 9 10—86 85 84 83 82

Library of Congress Cataloging in Publication Data

Schmidt, Terry.
  Managing your career success.

  Includes bibliographies and index.
  1. Engineers. 2. Scientists. I. Title.
TA157.S385   650.1′4   82-15364
ISBN 0-534-97948-3   AACR2

For Sinee and Cheryl

Short Table of Contents

Contents

9 Making Effective Career and Life Decisions, 105

Important decisions deserve time and thought. Define the problem and clarify your criteria. Let your intuition play a part too.

10 Defining and Achieving Your Goals, 115

Knowing where you want to go is necessary for good career management. Consider alternate routes to your goals. Use short-, medium- and long-term objectives to help you along the way.

11 Maintaining Success Throughout Life, 130

Your values and goals change as you grow and change. Recognize the signs that you are changing. Use your transition periods as opportunities for growth.

PART THREE   PUTTING YOUR PLANS INTO ACTION, 143

Be thinking about what you want for your next job. Use your contact network to find out what the job is all about. Then you'll be prepared if you decide to change or if you are laid off.

Begin by being more productive: work smart, not hard. Redesign your job to make it more challenging and satisfying.

The better you get along with others, the better you can do your job. Use every opportunity to improve your communications skills.

15 Learning for Your Job and for Yourself, 175

Analyze technological and company changes that could affect your job.
Use formal instruction, on-the-job training, and self-instruction to keep
current in your field and to enrich your personal life.

16 Putting It All Together, 189

Organize your career strategy, test it, improve it, and put it into action.

APPENDICES

Preface

Destiny is not a matter of chance, it is a matter of choice. It is not a thing to be waited for, it is a thing to be achieved.

William Jennings Bryan

WHO THIS BOOK IS FOR

This book is about you, your career, and your life. It is designed to guide you in developing effective strategies to achieve your career and life goals in a rapidly changing world.

Technical professionals and managers face unique career opportunities and risks which require special career management techniques. This book describes career management techniques developed with, tested by, and of proven value to engineers, scientists, computer experts, and others in technical fields. With these, you can formulate and implement flexible, realistic, and effective success strategies throughout your life.

Let me share three assumptions I make about you, the reader, and how these assumptions shape the approach I have taken: First, you are intelligent and curious. Because you have these traits, the book is interdisciplinary and synthesizes concepts drawn from the behavioral sciences, management sciences, and engineering and technical fields.

Second, while career success is important, other aspects of your life are also important. Because of this, I describe ways to define and reach your career objectives without neglecting the other factors vital to overall success and satisfaction.

Third, you are willing to manage your career. Thus I suggest ways to think strategically, develop sound plans, take action, and use your job to contribute to both the organization's and your own goals.

If these assumptions fit you, I've found the right reader and you've found the right book.

WHAT THIS BOOK DOES

Strategic career management puts you in control of your destiny. This is not a one-time planning event, but a lifelong process of planning, decision-making, and action to match your personal goals and skills with challenging opportunities in the work world. The process helps you to:

- get more out of your present job
- understand your motivations, skills, interests, and values
- increase current and future income potential
- make better career decisions and job choices
- achieve your work and life objectives
- avoid the career problems resulting from rapid change
- enjoy challenge, satisfaction, and growth at all life stages

Throughout the book, I cite actual examples of men and women who are facing issues which may be similar to yours. I also share my own career experience as an engineer, management consultant, entrepreneur, and career workshop trainer.

You will not find any magic formulas or foolproof success methods here—or in any book. But you *will* find practical principles, guidelines, and ideas to stimulate your discovery of methods that work in *your* life.

HOW TO USE THIS BOOK

First skim the book to pick up the main ideas, then read it more thoroughly.

To accommodate the busy professional with limited time for reading, chapters are short. Key points are summarized at the end of each chapter for later review. Questions are posed to help you relate the ideas to your own situation. Reference readings of particular value are cited.

This is not a textbook to be read once and shelved, but a practical career technology handbook to be called on many times in the future. Personalize this book: read with pencil or pen and mark pages to trigger your thinking as you review these concepts in the future. Keep a separate journal or notebook to record your ideas, plans, and progress.

I hope this book becomes part of your permanent collection of important reference handbooks. More important, I hope the ideas become part of your daily work and life.

ACKNOWLEDGEMENTS

Many people deserve credit for their help in shaping this book. Donald B. Miller and Arthur Glazer provided thoughtful critiques and valuable guidance from the start. My associates Jim Bogaty, Charles Brocatto, Alan Crosby, John Kizler, and Gary Lloyd gave helpful suggestions for the final manuscript.

Editorial assistant Peggy Soares untangled numerous unintelligible phrases. Typists Suzanne Crane and Monica Lee responded to tight deadlines with polished work. Sinee Schmidt gave me understanding and support when I needed it most. Publisher Alex Kugushev provided just the right mix of encouragement and prodding.

My sincere thanks to you all.

Terry D. Schmidt
Washington, D.C.

Managing Your Career Success

PART ONE

PLANNING YOUR CAREER FUTURE

CHAPTER 1

Defining What Success Means to You

I have learned that success is to be measured not so much by the position that one has reached in life as by the obstacles which he has overcome while trying to succeed.

Booker T. Washington

OVERVIEW

Career satisfaction and life success are universal goals, but few people have a coherent strategy for achieving these goals. This book is designed to improve your ability to develop a realistic, flexible, and effective success strategy. The starting point is to:

- Clarify the meaning of career and life success
- Identify the variables which affect that success
- Take charge of those variables you can control

YOUR PATH IS UNIQUE

We each chart a unique path in our career and life. Each person enjoys victories and suffers setbacks while striving to achieve lifelong challenge, growth, satisfaction, and success.

Defining Success

But what are satisfaction and success? The answers are as diverse as human beings are numerous. Comments by career workshop participants illustrate some of the interpretations:

"I really care more about *who* I am than *what* I am. The most important thing to me is who I *become*, in a qualitative sense of becoming. Success

BROOM-HILDA
by Russell Myers

Reprinted by permission of Tribunal Company Syndicate, Inc.

to me is defined in terms of integrity and growth, rather than position, status, power, or wealth. I would like persons who know me to say 'it was good to have known him.' If I can do that, I know I'm a success." (37-year-old mechanical engineer)

"My goal is to become the youngest president this company has ever had. I get my kicks from beating the competition and advancing up the company ladder. To me, success means moving as far and fast as possible." (26-year-old computer systems analyst)

"I've won my share of organization battles and reached a comfortable responsibility level. I get my primary fulfillment from watching my daughter grow, coaching the Little League team, and working with local volunteer groups. To me, that's a *real* challenge." (47-year-old R & D director)

"I feel successful when I'm working on an important technical problem with stimulating, competent colleagues. If I can help advance the state of the art, I'm content. I've been offered a management job, but it wouldn't make sense for me. I'm most satisfied doing what I'm doing now." (31-year-old polymer research scientist)

These four individuals defined success as inner growth, job advancement, family and community involvement, and technical contribution. We must each define success in our own terms.

Your own definition derives from the many things you do—you work, you play, you relax, you learn, you raise a family. Success takes on different meanings in different phases of life. Defining success requires understanding your values, needs, abilities, and aspirations. This is not achieved all at once, but over a lifetime of experience and growth.

Success in the 80's

I believe that the commonly-accepted career success definition of the 50's, 60's and 70's has become obsolete in the 80's. Rapid upward mobility was the norm through the early 70's, fueled by economic expansion and work force shortages. Individuals enjoyed rapid promotion and swiftly rose to higher echelons of position, responsibility, title, and salary.

But today's environment has changed—dramatically. The reality of slower economic growth, maturation of the baby-boom generation, an abundance of college-educated talent, more women in the professional work force, and a trend to later retirement mean fierce competition for promotions which would have been automatic a generation ago.

Individuals who define their success primarily by promotion and reaching high organization levels will experience unnecessary pain and conflict during the 80's. I know many people with the competence to advance to the top of their company or profession. In fact, I know far too many competent people; the sheer mathematical odds make it improbable for most. But I refuse to judge their career success—or to let others judge me—simply by the position attained. Fame, fortune, rank, and power are not required for a challenging and productive career.

More important indicators—and achievable for all people—are having fun at what you do, feeling that your work is meaningful, having an impact on what happens, developing a sense of personal worth, and helping your peers and subordinates achieve their goals.

I like the term "personal vitality" suggested by Donald B. Miller. (See the Recommended Readings at the end of this chapter.) Personal vitality means the desire, capacity, and power to perform effectively and vigorously in life and work.

Career satisfaction results from actively pursuing personal and professional excellence. This increases the chances for, but does not require, rapid upward mobility. In fact, it is better tactically to consider promotion and salary growth as acknowledging the success you have already achieved, rather than as the primary purpose of your strategy.

YOU HAVE MANY DIMENSIONS

At a cocktail party years ago someone challenged my thinking by asking the proverbial cocktail party question: "What do you do?" I answered by mentioning the job I then held, aviation policy analyst with the U.S. Department of Transportation. He asked, "Is that all?" I was proud of my job and insulted by his abrasive response.

He sensed that I was offended and explained: "I meant no offense; you missed the point of my question. You answered in terms of the job you hold. But you don't work 24 hours a day, and you won't have that job forever. What other things do you do, what other roles do you play in addition to your work role?"

This exchange provided me with fuel for later thinking about what gives me satisfaction. I realized that I am several different "me's" combined in one package. My professional self is concerned with job, work and career pursuits. But I have other important selves. I have a family self, with multiple roles as spouse, lover, father, son, and brother.

My spiritual self seeks to develop a life philosophy, understand the larger order of things, and know God. I have a social self, with friends, acquaintances, and memberships in civic and social organizations. My physical self enjoys jogging and plays mediocre tennis. I have a personal self that plays chess, attends theatre, putters around the house, and demands time alone.

We each perform multiple, simultaneous roles. Each role imposes its own demands; each provides satisfaction as well as frustration. The importance of each role changes with time. Each role—plus new ones yet to be discovered—is a potential source of enrichment and growth. The challenge is to synthesize these changing roles, and make life and work decisions which strike a satisfactory balance.

While work is not the sole factor in personal satisfaction, it is fundamental for the simple reason that you spend more time working than doing anything else in your life. Work unhappiness spills over into your personal life. Even if your primary satisfaction comes outside work, your job is still necessary for the income to pursue other interests. To get the most from your career, and to use your work to enhance overall life satisfaction requires that you examine and understand what influences your career progress.

IDENTIFY CAREER MANAGEMENT VARIABLES

Ironically, all technical professionals spend years acquiring their career talents but few actively *manage* their career progress. Some become aware of the need only when rudely awakened by a project cancellation or layoff. Others equate career management with job change, but fail to understand how each job fits into a lifelong pattern. Still others defer their critical career decisions to the company or to fate. The predictable result is a job mismatch and a dissatisfied individual who performs below his or her potential for career satisfaction and life success.

Strategic career management is the art and science of making decisions and taking actions to optimize your overall satisfaction. If we define *Y* as career satisfaction and success, the variables of career management can be portrayed with an equation:

$$Y = F \text{ [understanding and managing } (E + O + I + P + \ldots)]$$
where E = external factors, trends, and dynamics
$\quad\quad O$ = organization and industry factors
$\quad\quad I$ = interpersonal factors
$\quad\quad P$ = personal factors

The career strategist periodically solves the equation for an optimum value of *Y*. It must be solved periodically because the values of the variables constantly change. This is not a literal equation, but it does suggest the primary variables which affect your career. These variables are multiple, interactive, and dynamic; the career strategist makes appropriate plans based on understanding those variables.

The Primary Variables

External factors shape the domestic and world environment. Technological, political, economic, and social in nature, their complex interaction shapes the industries which provide work opportunities. Your actual *control* of these dynamics is limited. But you can—and should—observe and understand how these factors affect your career future. You'll want to monitor "leading indicators" which signal opportunity or risk for you, and be prepared to switch jobs, industries, geographic locations, and so forth. (Chapter 2 tackles this subject in more detail.)

Organization factors are company-internal variables which affect your career progress and day-to-day job satisfaction. Two are of special importance: the organization's "personality" and its competitive strategy. (Chapter 4 analyzes organization personality; strategy is discussed in several places.)

Each organization has a unique personality which determines how people within the organization act toward each other. Understanding this personality is the key to determining how the company fits your own needs. By understanding personality, you can make the most of opportunities in your company, create new ones, and use your present job as a springboard to progress inside the company or elsewhere.

The organization's competitive strategy has a major affect on your career progress. Organizational survival and success in the turbulent 80's demand sound strategy. The key issue to understand is the quality of that strategy and how you fit with it.

Interpersonal factors are crucial to career success and personal satisfaction. Too many people think that doing a good technical job is all that is needed for job success. But organizations are comprised of people, and virtually every aspect of your career depends on action by other people. Interpersonal skills are vital for moving up in an organization, influencing your boss, getting cooperation from colleagues, interacting with top management, changing jobs, and countless other situations in work and life. (Chapter 14 focuses on these issues.)

Personal factors entail your work personality, style preferences, strengths, and limitations. The key personal factors to understand are *what you do well and what you enjoy doing.* Each person has a unique mix of talents, skills, capabilities, knowledge, and interests. The single best nugget of career planning wisdom is to become aware of those, then find or create work environments which call for what you do best. (Personal factors are examined extensively in this book.)

Note that your control increases with each set of factors. Your influence on the external factors is slight; you have some impact on organization factors. You have more influence over interpersonal factors, and the greatest influence on personal factors.

Career management means *understanding* the variables affecting you and *controlling* those you can. This entails *identifying* these factors, *interpreting* their affect on you, *defining* your own life needs and goals, *developing* an appropriate career strategy and plans, and *implementing* your plans to make your life dreams come true. In a paragraph, that's what this book is all about.

SUMMARY OF KEY POINTS

- Success is defined by your own criteria. Enrich your definition to recognize that success is not measured primarily by visible external achievement, but through pursuit of personal excellence.

- Experiencing success is an active, ongoing process of finding challenge, satisfaction, and fulfillment throughout your life. Self-discovery, growth, and learning are the heart of the process.

- Because career is an integral component of life, your career must be planned and managed within the context of your overall life needs, objectives, and concerns.

- The purpose of this book is to help you meaningfully define career and life success, understand the many variables affecting success, and formulate and implement realistic strategies to better achieve that success.

QUESTIONS TO CONSIDER

1   How do you currently define career and life success? List some measures.

2   How do those people you respect or love (your mate, parent, friend, boss, etc.) define career and life success?

3   What are the major roles you perform (e.g., parent, gardener, friend, bridge player)? Which give you the greatest satisfaction? Which demand the most time?

4   Compared with other aspects of your life, how important is your career? Ask yourself some broad questions about life balance.

5   What "return on investment" do you expect and need from your job?

6   What is the one aspect of your present job which excites you the most? Which disturbs you?

7   What do you hope to achieve from reading this book?

RECOMMENDED READINGS

Note: The references below and at the end of each chapter are listed in the order of their value to the reader.

Miller, Donald B. *Personal Vitality*. Reading, Massachusetts: Addison-Wesley Publishing Co., Inc., 1977. A stimulating book about the ability of individuals, organizations, and nations to adapt to new information and changing world conditions. Includes many interesting personal growth projects. *Personal Vitality Workbook: A Personal Inventory and Planning Guide* is a companion volume with individual planning and goal-setting exercises.

Kiev, Ari. *A Strategy for Success*. New York: Macmillan Publishing Co., Inc., 1977. An upbeat book which makes you think. Especially good treatment of how to change unwanted habits and focus on your important objectives.

CHAPTER 2

Recognizing Signs of Opportunity and Risk

My interest is in the future, because I am going to spend the rest of my life there.

Charles Kettering

OVERVIEW

The rapidly changing environment of the 1980's has important consequences for your career. Thus strategic career management begins by observing the change dynamics in our country and the world, and interpreting how such changes may affect your career. You can improve your career strategy by understanding:

- Technological, economic, political, demographic, social and cultural change factors
- How to analyze change factors which affect your industry, company, and specialty
- The effect of specialization and employment volatility on your career
- The impact of changing social norms, values, and demographics on work in the future
- The career problems and opportunities such changes create for you

SCAN THE CHANGING ENVIRONMENT

Looking at the 80's and beyond, one thing is certain: accelerating change. The effects of change are visible everywhere and influence every aspect of our work and life.

Change generates both risk and opportunity. Learning to understand change—even to thrive on change—may be the most important career management skill. The more you understand basic change factors, the better you can manage their impact on your career.

Cultivate the habit of "scanning the environment." This is not an academic exercise, but a practical analysis to determine how your career future may be affected by the complex interaction of scientific, technical, economic, political, demographic, cultural, and social factors.

As a starting point for your analysis, Figure 2–1 outlines major international and domestic change factors. Not all of these equally affect your future career. The purpose of scanning the environment is to identify issues with a *dominant* impact on your industry, company, profession, and specialty.

Figure 2–1   Major forces for change in world and
United States environment

World social/physical system

　Accelerating development of global interdependence

　Erosion of traditional Western political and military alliances

　Growing Soviet military strength; potential for direct conflict

　Global energy crisis; dependence on Middle East oil

　International economic volatility, speculation, currency shifts

　Global food shortage and continued population growth in Third World

　Increasing income gaps between industrialized and developing nations

　Political/military instabilities; nuclear proliferation; terrorism

　Decline in "limitless" resources

　Growth of multinational corporations as potent international political force

U.S. society of the 1980's

Technological/scientific

　Continued growth of technology; secondary consequences like pollution, stress

　Unanticipated "secondary effects" of innovation (e.g., Three Mile Island)

　Spiraling state of the art in virtually all technologies, hastening skills obsolescence in some specialties

　Emergence of such new technologies as biomedicine, robotics, genetic engineering

　Explosion in information technology, computer applications

　Increased foreign competition in areas of traditional U.S. dominance

International technology transfers

Trends to renewable technologies

Economic/industrial

Persistent inflation, capital shortage, dollar instability

Declining productivity of U.S. labor force, industry

Critical dependence on imported oil

Continued increases in costs of housing, energy, health, food

Structural changes in industry; shift from a production-based economy to a service and information economy

Aging physical infrastructure—decaying roads, bridges, sewage systems

Systemic problems in steel, auto, other capital-intensive industries

Escalating government spending, persistent budget deficits

Political/governmental

Competing demands on limited financial resources

Decreasing confidence in government, political leaders, efficacy of the "system"

Splintering of traditonal party system

Emerging dominance of "single issue groups" supporting/opposing candidates and issues

Increased business and special interest groups' activism in political process

Increasing problems for society in making strategic/quality choices

Demographic/family

Changes in labor force by age/sex/race

Rapidly aging U.S. population, growth rate declining toward zero population growth

Population redistribution to western and southern "Sunbelt" states

Emergence of dual career and two-paycheck families, single-parent and extended families

Declining cohesion, support roles of family for its members

Cultural/social

Increasing unemployment, growth of "underclass" without marketable skills

Rising expectations for quality of life, work satisfaction

Increasing educational levels of work force; shortage of jobs requiring college education

Decreased ratio of number of wage earners to number of retired people

Legitimacy crisis of traditional institutions and authority systems

Increases in leisure time, disposable income

Desire for independence, self-expression, participation in decision making on the job

The following sections explore some major factors affecting us all, and suggest issues for you to examine in further detail.

ANALYZE HOW CHANGE AFFECTS YOUR INDUSTRY/ORGANIZATION

Technological and other changes affect every industry and company through process or product *substitution.*

To cite a nostalgic example of product substitution, consider what happened to a certain instrument comprised of a ruler and medial slide, graduated with similar logarithmic scales and corresponding antilogs used for rapid calculation—the slide rule. A decade ago, the computational workhorse of engineers and scientists was a multi-scaled, leather-holstered "slip-stick." But the technological miracle of the hand-held calculator made the slide rule obsolete. Today, these relics collect dust in lower desk drawers. Slide rule production ceased in the mid-70's; the industry no longer exists.

To explore change factors affecting you, first analyze changes in your industry, then examine your organization's strategy for responding to those changes.

Industry Analysis

The success of the industry you work for has a major effect on your prospects in that industry. Thus it is important to examine industry growth rate, economic health, future threats and opportunities, and factors generating demand for its products or services.

Industry Life Cycle

The well-known life-cycle curve in Figure 2–2 describes the growth profile of most industries (as well as sub-industries, companies, and individual products). Industry growth follows a predictable four-stage cycle of emergence, growth, maturity, and decline. The time required to traverse the cycle varies; major industries span many decades, others last just a few years.

Career opportunities tend to be better in emerging and growing industries—robotics, genetics, data communications, and microcomputers, to name just a few. Growth creates additional openings, higher salaries, and better promotion prospects. Mature industries—steel, rubber, aircraft, machine tools, heavy manufacturing—offer the bulk of jobs but fewer growth opportunities. Avoid (or be prepared to leave) those in serious decline (automobiles, railroads) if your career has many years ahead.

Figure 2–2 Life-cycle curve

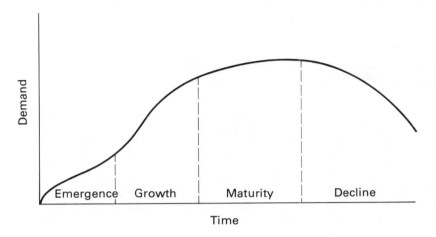

Discretion is needed in defining "industry." Large industries are comprised of smaller industry segments with varying growth rates. In the computer industry, for example, mainframe computers are in the maturity stage, while minis and micros are in rapid growth.

Industry growth does not mean uniform demand increases for all occupations in the field. The U.S. Department of Labor predicts that between 1978 and 1990, for example, demand for computer programmers will increase by 74–87%; for computer operators by 88–101%; and for EDP machine mechanics by 148–175%. But key punch technician jobs will actually decline as optical reading techniques replace manual entry.

Analyze Growth Trends

Examine the underlying causes behind industry or product growth trends to determine their stability, strength, and long-term prospects. High initial growth technologies can easily fizzle (the video disc); some slow-growth industries have bright futures (solar applications).

Cultivate the skill of tracing the reasons for growth (or lack thereof) to causal factors. These are usually a combination of social, economic, cultural, technological, demographic, and political factors.

Consider a few diverse and simplified examples of tracing product or industry growth to causal factors. Microwave oven and convenience food sales will climb in the 80's; the increase in working women and single-parent homes means less time for traditional cooking. The home video game and cable television industry will grow; congested streets and fear of crime encourage people to entertain themselves at home.

The recreational vehicle industry will remain stagnant due to the cost of gasoline.

Career Strategy in a Stagnant Industry

Persons belonging to (or able to join) growth industries have an enviable career planning problem—how to choose from multiple opportunities. The more serious problem is for those in stagnant or declining industries.

Take the nuclear industry, born with high hopes and rapid expansion. Despite the energy shortage, its future prospects are gloomy due to public fears of catastrophe kindled by Three Mile Island. Bruce Charles, a 35-year-old research metallurgist with a major nuclear plant manufacturing company, described his career strategy as follows.

> "I don't see us building any new plants during the 80's. Despite the industry problems, I like the company and my job. I do research in strengthening ceramic and metal materials in large castings to minimize molecular-level grain boundaries. These defects are the leading cause of large machinery failure, and that problem is not unique to the nuclear industry.
>
> "I hedge against career risk by going to technical conferences, not just nuclear industry conferences, but professional association meetings attended by other ceramics and metallurgical engineers. I meet lots of new people. We exchange business cards and literature. This gives me a good way to find out what's going on in other industries.
>
> "I try to stay aware of new metallurgic techniques such as the plasma process, a hot new area. If I want to change industries someday, I can start with my contacts doing similar work in totally different industries. My industry may be in trouble, but my specialty is not."

Clouds on the industry horizon don't necessarily mean you should abandon the field. But you should be prepared for mobility across industries should you want or have to change. Charles keeps abreast of applications of his specialty and maintains contact with people in other industries. His strategy is sound.

Analyze Your Own Industry

Here are some questions to ask of your industry:

In what life-cycle stage is the industry?

What are the major growth segments of the industry?

What future technological developments are likely to change the industry? What threats are posed by substitute products?

What environmental dynamics (social, political, economic, cultural, demographic) most affect it?

What trends provide the industry with its best future opportunities?

What are the greatest risks faced by the industry?

How strong are international competitors?

In what areas does the U.S. hold a technological edge over foreign competitors?

The United States is shifting from a manufacturing to an information-based economy. Expect a continuing decline in labor-intensive manufacturing industries; the lower labor costs of such Third World countries as Taiwan, Korea, and Brazil shift production economies in their favor. Products with a high labor component relative to the technological input suffer the greatest threat.

The Productivity Challenge

The biggest challenge facing American industry in the 80's is reversing the productivity decline. Declining investments in technological innovation and focus on short-term profits during the past two decades have eroded the competitive edge which the United States long enjoyed in virtually every field.

Foreign competition is not limited to computers, television, electronics, telecommunications equipment, industrial robots, and other high-tech fields; one-third of America's major league baseball players step to the plate swinging Japanese bats.

The productivity challenge has some direct personal implications. Job security and advancement in your industry is dependent on improving industry economic health and productivity. Marginally productive or inefficient employees will be trimmed from the payroll. The career edge goes to individuals able to innovate, cut waste, improve value, do more at lower cost, increase efficiency, and make their work environment more productive.

Organization Analysis

Growth industries don't always mean success for your firm; mature industries needn't spell disaster. Because some firms will thrive and others won't survive whatever the general industry fortunes, the critical issue is whether your organization has a coherent strategy for the future.

Develop the habit of asking and getting answers to basic organization strategy questions. Ask the following questions of your current organization and any new ones you may consider joining.

Where does the company plan to be five years from now (in terms of product line, technologies, sales, market position, and reputation)?

What is it doing to get there?

What are its key strengths? Weaknesses?

What is its track record in anticipating and reacting to trends which affect the business?

How much does the organization invest in R & D (compared with industry norms)?

What are its most innovative characteristics?

How is the competitive strategy developed or updated? How is it translated into plans and action?

How well is the company managed?

Equally important, what skills and knowledge do you bring to help the company implement its strategy? Do areas of future business emphasis coincide with your own interests? What contributions can you make towards key organization objectives? Are you excited about your future in that company?

Your success is linked to the success of organizations you work for, and organization success demands coherent strategy. The absence of clear strategy requires that you plan your career in a vacuum, or plan your way out of that organization.

Impact of Government Policies on Industry

Keep your eye on Uncle Sam. Federal aerospace and defense programs, directly or indirectly, employ nearly half the nation's engineers and scientists. Federal policy, R & D funding, and system procurement decisions stimulate demand for some specialties while creating oversupply in others.

President Kennedy's 1961 commitment to land Americans on the moon stimulated technology and built a new cadre of scientific and engineering talent. President Nixon's 1970 creation of the Environmental Protection Agency spawned the pollution control industry and a new breed of technologist, the environmental engineer. President Reagan's 1982 decision to beef up the military created thousands of new jobs in aircraft, munitions, computer, and other defense related industries.

NASA, EPA, and Defense and Energy department programs have the greatest impact on technical employment. R & D funding patterns are the most important leading indicators of government policy directions. The National Science Foundation keeps close track of the magnitude and composition of federal spending. Trade association and professional society newsletters also monitor government trends which affect your industry or profession.

REDUCE TECHNICAL EMPLOYMENT RISKS

The success of the American industrial system depends on a high degree of individual specialization. Yet the specialization necessary for organization excellence creates individual risks. The skills and knowledge of your specialization may become obsolete. You may be vulnerable to layoffs arising from shifts in technological emphasis or market demand.

Watch for Signs of Industry Trouble

The risk is greatest for those who narrowly restrict themselves and ignore the signals of waning demand for their specialty. An example of how myopia creates career problems: A NASA physicist responsible for the Saturn V first-stage thrust valve knew absolutely everything about that particular piece of hardware. It was obvious that after the Apollo program, there would be little demand for those skills in NASA and even less in the private sector. But he gave little heed. His layoff was a brutal shock and he floundered for a year before his career got back on track. Another NASA scientist working on the Apollo program planned ahead by thinking about what he wanted to do next and exploring outside opportunities. He made a smooth transition to a university research position.

The signs that your industry or specialty is troubled usually appear in sufficient time to prepare, but you must react to them.

Be Flexible

Employment prospects are bright for scientists and engineers during the 80's. But while overall demand is strong, there will be demand/supply imbalances in sub-specialties. Thus it is important to remain relatively flexible and prepared to switch to related fields when necessary. For example, there is a surplus of mathematicians and a shortage of computer systems analysts. Mathematicians unable to find jobs traditional to their field can become computer analysts without difficulty. Don't unnecessarily restrict yourself.

Employment demand is highly sensitive to economic cycles. Industry employment hiring patterns follow economic swings, adding workers during favorable conditions and cutting back in stagnant periods. Figure 2–3 traces some industry hiring patterns from 1961 to 1981. This Deutsch, Shea, & Evans High Technology Recruitment Index was compiled from employment advertising in technical journals, and from newspaper classified and display advertisements.

Marginal performers are first on the layoff list and most vulnerable to economic downturns. Companies often replace the more experienced

Figure 2-3 Volatility of scientific/engineering
employment demand

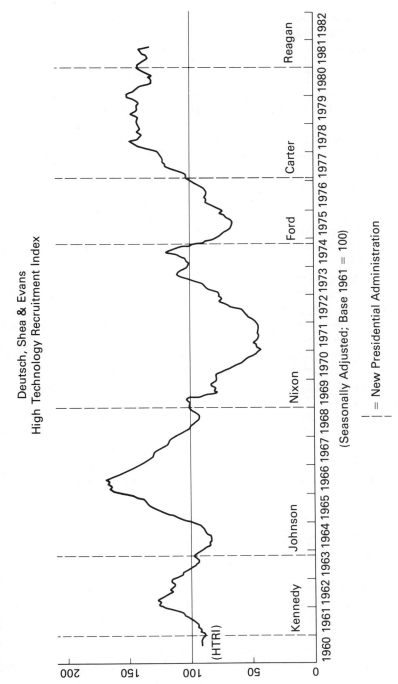

Deutsch, Shea & Evans
High Technology Recruitment Index

(Seasonally Adjusted; Base 1961 = 100)

= New Presidential Administration

Reprinted by permission of Deutsch, Shea, & Evans, Inc., New York.

(often spelled "expensive") employees with new graduates who are technically current and more willing to relocate.

Minimize the Risks, Maximize the Opportunities

You can minimize the overspecialization and employment volatility threat in three ways. First, ensure solid performance: deliver what you are paid for, plus a bit more. Second, maintain your skills through lifelong continuing education. Third, stay aware of trends which indicate future demand downturns in your field and prepare for cross-field mobility.

The spectrum of individual specialties required in today's large, complex scientific and technical organizations creates personal opportunities as well as risks. The trend to technical specialization requires a new breed of technologist, the *integrator*, to organize different specialties into a coherent whole. Such persons speak the many languages of specialized discipline and apply the expertise of diverse individuals to important organization objectives.

The integrator isn't a position on an organization chart. Rather, integrators are people who look beyond the confines of their specialties and ask more intelligent questions. They are attuned to the changing world, and identify emerging needs to which their company can respond. Specialists turned generalists, they convert their understanding of technology into internal projects and ideas. They are skilled at motivating and getting cooperation from others through project management and matrix methods.

The proliferation of methods and techniques creates an opportunity for you to "hybridize"—add a new skills area to your existing knowledge base. For instance, geologists who learn satellite techniques for locating mineral and hydrocarbon deposits can hang up their rock hammers in favor of satellite imagery interpretation, thus enhancing their careers.

DEMOGRAPHIC AND SOCIAL CHANGES TO EXPECT

Every person who will be in the work force during the next twenty years is alive today. Demographic and social change factors, even more than technological change, will dramatically shape your work environment of the 80's.

You didn't choose when to be born, but your birthdate affects your career promotion prospects. If you were born before 1945, your chances of rising are better than if you were born after this date. It's a matter of simple demographics.

Demographic Changes

During the baby boom which followed World War II, the birth rate soared from 2 to 4 million per year, maintaining high levels until the mid-60's "baby bust." The baby-boom generation will reach their mid-career years during the 80's. In this decade, the 35- to 44-year-old population group will increase from 25 to 36 million, a 42% jump, versus only a 21% growth in the number of mid-level jobs.

Disillusionment awaits many baby-boomers whose career expectations were shaped by the rapid advancement of older friends and siblings. Scientists and engineers have an advantage over their non-technical brethren, as technical progress will create new jobs and work force shortages. But the best jobs will be far fewer than the number of people competing for them.

If you are part of this group born after 1945, develop a view of career success not contingent on frequent promotion, and create opportunities for challenge and satisfaction off the job as well as on.

Those born before 1945 and now in their late 30s and beyond will feel a different demographic pinch: greater competition for their jobs from ambitious baby-boomers. Because authority now corresponds more with knowledge and experience rather than age and seniority, they will experience a pressure to perform or be overtaken by ambitious up-and-comers.

Social Changes

More subtle but powerful changes are occurring as work values shift away from the traditional ethic. The carrot-stick motivation formulas used a generation ago don't work today. Carrots have lost their appeal; sticks don't carry the same punch. Anyone whose work requires cooperation from others—and that's just about everybody—must recognize the new work values, re-examine his or her own values, and use new methods to motivate increasingly independent workers.

Alan Williams grew up during the depression. Following distinguished service in World War II, he earned a mechanical engineering degree and joined a consumer appliance manufacturing firm, where he is today. He describes his experience in adjusting to the changing values of the workforce:

> "When I became an engineering supervisor I didn't realize the toughest part of my job would be dealing with people. It's been a challenge and a battle. During the 50's, the people I hired shared my values and goals. Managing was easy—I told people what needed to be done, and they did it.
>
> "In the 60's a new breed started coming aboard. They looked different, they acted different. I guess the Beatles were responsible for long-haired

men. At first I wouldn't hire a longhair, but later I couldn't find enough short-haired engineers. Some engineers on my staff even grew long hair, but I couldn't make them cut it any more than I could get my teen-age boys to trim theirs.

"But the hair thing was less traumatic than the authority issue. The younger generation questioned everything. They'd challenge my authority and expertise. They demanded a role in setting priorities and making decisions—unheard of when I started working. The Viet Nam War was going on, and I had some heated battles with those opposed to the war. Funny, but over time I started to agree with them.

"In the 70's it was the women, and women's liberation. We hired more female engineers and the secretaries wanted to do more than type. Even my wife, who hadn't worked in 20 years, started a career. I was openly called a male chauvinist pig. My attitudes are different today, and some of my best engineers are women.

"I'm not sure what to expect in the 80's. But I know one thing: I can't just assume that other people see the world as I do. The process of accommodating different value systems is painful, but it's necessary for the company and may even be healthy."

Employees today bring new personal values to the job. Employer loyalty is declining as increasingly mobile workers switch companies with ease. Today's employees place less importance on work and more on leisure, family, and personal pursuits than did their predecessors.

People want more meaningful work where they can have an impact. They demand greater on-the-job autonomy, self-expression, and participation in decision making. Progressive organizations respond to the needs of individuals and dual-career families through greater use of flex-time, part-time work, shared jobs, job rotation, and other methods. Supervisors and managers face the challenge of creating work environments where individual expectations can be met while also achieving essential organization objectives.

Technology of the Future

The new technology of microcomputers and the paperless offices has far-ranging implications. Many companies now equip employees with a home computer terminal; it may not be necessary for you to commute to work in the future. How would working primarily from your home affect your life? What new possibilities for restructuring your family, personal, and work life-style would this flexibility provide you? What are the implications of reduced on-the-job personal relationships for your work satisfaction? These are exciting questions to think about; the day when you must answer them may not be far ahead.

Change is the one common denominator of the 80's. No one can accurately predict how the future will unfold. But by observing and understanding technical, economic, political, social, and cultural change factors, you can better manage change's impact on your career. Prepare now to minimize the risks and take advantage of the career opportunities change creates for you.

SUMMARY OF KEY POINTS

- The purpose of "scanning the environment" is to improve your career strategy. Plan with *commitment* for what is certain; with *contingency* for what is uncertain; be prepared to *react* to the unknown.
- Since your career growth depends in part on the future direction, growth, and success of your employing organization, pay close attention to your organization's strategy.
- Don't overspecialize. Be prepared for cross-field mobility should you want or have to change.
- Changing work values challenge us to create work climates to accommodate diverse personal motivations and work styles.
- Lifetime learning is a necessary part of career strategy. To keep up is difficult, but merely to maintain the status quo invites obsolescence.
- The future belongs to those who sense the opportunities change creates and prepare themselves with the skills and knowledge needed to meet tomorrow's demands.

QUESTIONS TO CONSIDER

1   What basic environmental trends and dynamics most affect your industry?
2   To what type of external events is your profession most sensitive or vulnerable?
3   What changes do you foresee over the next ten years to your industry? What are the implications for your skills?
4   Does your organization have a coherent strategy for the future?
5   What degree of cross-field mobility do you currently enjoy? If you were to be laid off tomorrow, what would you do?
6   How well prepared are you to change work activities and interests, to switch fields or industries? Are you in danger of being too spe-

cialized (either in reality or the perception of others) to adapt to
new situations?

7   What effects of changing social values do you see in your organization?

RECOMMENDED READINGS

Toffler, Alvin. *The Third Wave.* New York: William Morrow & Co., Inc., 1980.
This sweeping synthesis of the present and future provides a provocative
look at the 80's and beyond. Toffler's earlier book, *Future Shock*, is also
worthwhile.

Porter, Michael E. *Competitive Strategy: Techniques for Analyzing Industries and
Competitors.* New York: Free Press, 1980. A comprehensive treatment of
analytical techniques for analyzing the competitive dynamics of industry.
Helpful in understanding the competitive issues affecting your firm or
industry.

Mesthene, Emmanuel G. *Technological Change: Its Impact on Man and Society.*
Cambridge, Massachusetts: Harvard University Press, 1970. In this brief
book, the Director of Harvard's Program on Technology and Society outlines the effects of technology on social institutions and values. Includes
annotated bibliography of major works on the topic.

CHAPTER 3

Managing "You, Inc."

To do a common thing uncommonly well brings success.

Henry J. Heinz

OVERVIEW

You have all the resources needed to reach your career objectives, but you must use them effectively. To better use these career resources, it helps to view them as a system to be managed for growth, satisfaction, and progress.

This chapter examines your career enterprise, and describes how to:

- Clarify the "invisible agreements" you make
- Make the most of the career resources available to you
- Concentrate on the career success factors you can control
- Manage your interrelated career functions in a businesslike manner

REVIEW YOUR PSYCHOLOGICAL CONTRACTS

How important is your work compared with personal and family concerns? What do you expect your company to give in return for what you contribute? We all answer these questions, implicitly if not explicitly. We answer them by establishing unwritten "psychological contracts" with ourselves, our families, and our organizations.

Written or not, they are important career agreements. You would be wise to review them from time to time because as conditions change, they may need redefinition.

Contract with Your Organization

Your "psychological contract" with your organization was established the day you took the job. It expresses what you will contribute—skills, energy, dedication—in exchange for what you will receive—challenging work, financial rewards, and the like. But in time, the terms of the original contract may become fuzzy as conditions change on both sides. Your present boss may not be the one who hired you. Your initial expectations may no longer be possible to fulfill; the organization's expectations from you may not have materialized.

Consciously or unconsciously, people adjust their work performance and attitudes to the terms of their psychological contract. If the job promised—and provides—continued challenge and enrichment, you are likely to invest more effort to hold up your end of the bargain. But if the company doesn't deliver as you expected, your performance and attitudes are apt to change. Like any agreement, your organization contract can be renegotiated when changes in the company and/or you unbalance the original terms.

You have three basic options if your agreement no longer works for you: change the situation, change your expectations, or change jobs. More often than not, you can redesign your job to make it more satisfying. But this requires definition of the changes you would like and the willingness to initiate a renegotiation. (*How* to do this is covered later.) Failing that, you can adjust your expectations to accept the status quo; however, if the required adjustment is too great, the results may still be unsatisfactory. Then you must consider the third option—find a more suitable work situation.

Contract with Yourself and Others

Your unwritten career agreement with yourself clarifies the investment you will make in your career, the importance of your career goals in relation to other concerns, and the trade-offs you will make. It expresses what you will and will not do to achieve your career objectives. This agreement boils down to many practical decisions. How much of your discretionary time and money will you invest in activities which further your career? How much work are you willing to carry home regularly? Would you cancel a long planned and much deserved weekend at the beach to work on a crash project when given only a day's notice?

Your third unwritten contract is with "important others" in your life—spouse, children, and others with legitimate claims on your time.

How much out-of-town travel, which keeps you from your family, will you accept? At what point do you say "no" to work demands and "yes" to the demands of others? How do you resolve the practical issues of household management and child care when your spouse also works? How do you reach solutions in which both of your career aspirations can be met?

The nature of all these contracts—organizational, personal, and interpersonal—changes over time, whether or not you are aware of it. Periodically review the terms of your unwritten agreements, especially when you feel discontent. Discuss with your boss or your family the contract terms to redefine an agreement which works for all parties.

UNDERSTAND WHO YOU WORK FOR

However humble or high, your present work is the starting point for all future progress. But you must understand who you work for—the answer may be a surprise.

You work for *yourself*. Regardless of who *employs* you, you are ultimately a one-person mini-conglomerate producing professional services for sale to an organization customer. The company purchases your skills, knowledge, and experience, paying with financial and psychic rewards. You work for yourself; the company is no more than your current customer. And there are many other potential customers—that's what job change is all about.

The implications of working for yourself are powerful: you are responsible for your own success. This concept threatens some people because it forces the issue of personal responsibility. Many are more comfortable nestling in the bosom of company security. If that appeals to you, fine. But beware of yielding psychological control of your career destiny to the company or to fate, or you may feel "stuck" and believe you have few options. You always have options, whether or not you choose to recognize or use them.

The organization structure of "You, Inc." is shown in Figure 3–1. Your one-person enterprise requires an intelligent, businesslike management approach, just like organizations with hundreds or thousands of people. As each department is discussed, ask yourself, as chief executive officer, how you can make the departments work together effectively to make your enterprise succeed.

Board of Directors

Who serves on your board of directors? What role do others play in making important decisions about your career? The members of your

Figure 3–1 "You, Inc." organization structure

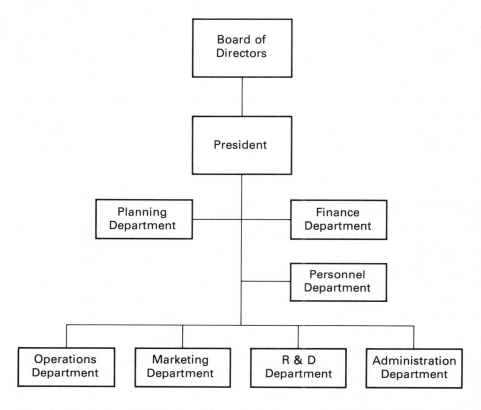

board are up to you. They may include your spouse, boss, colleagues, friends, and others of your choosing.

But your board may also include some people who no longer belong there. A good friend of mine suffered four years earning his doctorate. He confessed to me one day that he despised graduate school. Why then, I asked, are you working on a doctorate? He answered: "Both my parents have doctorates and they want me to have one too." So, to make them happy, he earned a Ph.D.—not for himself, but for them.

The degree of influence which board members exert over your enterprise is up to you. Remember that you are chairman of the board, and that while the opinions of others are valuable, you are ultimately responsible for your work and life decisions. Periodically review your board membership to see what changes should be made.

Customers

Management expert Peter Drucker succinctly defines the task of any business: *to find customers and satisfy their needs.* "You, Inc." serves many customers—your peers, management, others in the company, outside clients. But not all your customers warrant equal attention.

One question is crucial: how do your key customers measure success? Your key customers are the boss and top management of the department; they measure success as *achieving organization business objectives*. The effective strategist satisfies customer needs by focusing on those same objectives. Typical employees see the tree but not the forest. They perform their assigned job tasks, but fail to adjust those tasks to better contribute to key organization objectives. You gain a career advantage doing your job in a way that achieves objectives important to your key customers. In doing so, you enhance your career enterprise by generating additional customer demand.

Operations

This department "manufactures" professional services or products to meet customer needs. Managing this department effectively requires that you identify what your organization needs and choose from the many services you *can* provide to deliver efficiently those the organization most needs.

Focus your efforts on key objectives of your work unit, department, and organization. This may mean going beyond your written responsibilities to perform additional tasks critical to success, regardless of whether or not they fit your job description. Often you must creatively redefine your job to meet changing customer needs as organization requirements change.

Like any business, you have competition. You are not the only potential supplier; numerous other individuals within and outside the organization provide similar services. Your quality control function ensures that your services are as good as—or better than—those of the competition. This does not mean undercutting your colleagues but carefully analyzing customer needs and working with your colleagues to achieve organization objectives of mutual concern.

Marketing

You must continually market your services to current customers and to future prospective buyers. Many technical people consider the concept of personal marketing undignified, erroneously picturing the hustling jive of a used car or vacuum cleaner salesman. But this misinterprets both the purpose and method of marketing.

The purpose of marketing is to give you greater choice of future assignments. Most technical work is project-based, involving a series of sequential, temporary projects. Effective personal marketing creates a wider demand for your services and increases your options.

The best marketing method is to perform your current assignments well, gain visibility and demonstrate the capacity to handle greater responsibilities.

Marketing means staying active and visible inside and outside the company. Internally, it means becoming known to the organization "power structure" which controls reward decisions. Externally, it means acquiring personal contacts in other organizations and building a wider reputation to help you change jobs when you wish.

Research and Development

Your customer needs change as technologies evolve and the industry matures. To remain competitive, you must continually upgrade your product line. This is the job of your R & D department—to develop present and future marketable skills.

Personal R & D means lifelong learning to stay abreast of change. Much of this comes on the job, but it may warrant investing your own time and resources to develop a "distinctive edge."

Define your R & D program by examining the changes occurring in the industry/company, and then defining the "new services" you want to offer. New skills development supports the job you now have and prepares you for future opportunities. (Chapter 15 goes into detail on this important topic.)

Administration

This department is your "corporate memory," which documents your contributions by maintaining copies of all your work products, reports, memos, and so forth. It is responsible for coping with paperwork demands, maintaining reference files, names and addresses, and performing miscellaneous "housekeeping" tasks to stay organized.

Keep a work log. Document your most important achievements, what you observe, what you learn. Update it at least monthly. Awareness of your achievements helps when updating your résumé and can pay handsome dividends at appraisal time.

Staff Offices

In addition to these departments, your enterprise includes three important staff offices. Your *planning* function keeps abreast of the ever-changing industrial marketplace. *Finance* monitors your financial objectives to ensure you receive adequate compensation. *Personnel* manages your interactions with other people. (Chapter 2 described planning; Chapters 5 and 14 discuss finance and personnel topics.)

Periodically appraise how well each department and office functions. A candid review may identify specific ways to strengthen the performance of your career enterprise.

USE ALL YOUR CAREER RESOURCES

Get the most from your many career resources. Continually develop these resources; don't wait until you want to make a job change. This section gives some practical tips for developing and using each.

Your Current Job

Your best career resource is the job you now hold: all future options begin from the present. Ideally, your job lets you build new skills while providing satisfaction from solving challenging problems. To achieve these multiple objectives may require job redesign.

No job description is fixed in concrete—job scopes can be redesigned to the advantage of both you and the company. You can—and should—periodically renegotiate your job scope to create *new* areas for growth, challenge, learning, and development. You can recommend changes by understanding key organization objectives and evaluating your current responsibilities.

Your primary responsibility, of course, is to perform your present job well. But you can also use your current work to prepare you for future responsibilities. By keeping in mind several objectives for your next job and analyzing their skills and knowledge requirements, you can use your present job to develop these capabilities. Later I'll say more on how to revise your present job to prepare for the future.

Your Supervisor

Next to spouses, supervisors do more than anyone else to make your day stimulating, depressing, or in between. But unlike spouses, you can't choose supervisors—they come with the territory. Since you have a boss anyway, it makes good sense to use him or her to further your career objectives.

Most people make too little use of their bosses. Too often the supervisor is someone to be avoided, for various reasons: "He's not interested in my career." "She has no leverage." "He's a total loser." "We don't work well together."

This is a tactical mistake. Your supervisor can hinder or help your progress; you want his or her support. You don't have to be good friends. But you do need to develop a relationship of mutual respect and trust.

Figure 3–2 Use all your career resources

The best way to achieve a mutually supportive relationship is to *make the boss look good.* Regardless of your personal feelings toward the boss, it's in your best interest to help your boss succeed. This enhances your "clout" in renegotiating your job, choosing future assignments, and getting your supervisor to use influence on your behalf.

Your Mentor

Find a mentor, or let one find you. Mentors are more experienced individuals who give you career guidance and support. Former or current bosses make good mentors. So do colleagues, personal friends, and others who know you and are willing to assist.

Assistance from mentors depends on their experience, position, and contacts. Their most useful guidance is constructive criticism and suggestions. Mentors higher in the company can put in good words on your behalf, alert you to opportunities, provide introductions, pull strings, and explain the informal organization rules.

Mentors make excellent partners for your "development contract"—the personal learning and development plan strategists undertake. They can suggest development methods, help monitor your progress, listen to your ideas, provide candid feedback, and give useful advice throughout your career.

The mentor-protégé relationship is mutual respect, friendship, and trust. It's not a matter of manipulating another person. You have an obligation to maintain constant communication, not just call when you need a favor.

To find a mentor, stay active and visible on important projects and one will discover you. Being a mentor yourself can also be a very rewarding role.

Organization Career Resources

Most organizations offer useful services to the career strategist—training programs, counseling, printed material, and the like. Figure 3–3 lists some career resources your company may have. Find out more about what's available where you work.

Figure 3–3 Organization career resources

Strategic plans

Organization charts

Detailed department descriptions

Training and development programs

Management and/or employee development center

Human resource information system

Job opportunity forecasting system

Career paths/ladders

Job posting system

Upgrade and transfer program

Out-placement program

Tuition assistance programs

Manpower pooling and succession planning

Affirmative action programs

Job descriptions and related skill profiles

Performance appraisal system

Career Counseling

Assessment centers

Psychological counseling

Career development training programs

Skills and interests testing

Lifetime Learning Resources

Little elaboration is needed on the importance of lifetime learning, especially in rapidly changing professions. Books, libraries, periodicals, university courses, public seminars, and self-designed learning programs are your major resources.

Your Informal Contact Network

Your associates throughout the industry and organization comprise an informal network of information and support in your current job. They are an excellent information source concerning developments and opportunities elsewhere in the company or industry, invaluable when considering job changes.

Get to know as many people in the company as possible. Make sure your internal contact network includes people in other functions and departments. Everybody you meet—whether in the cafeteria line or a task force meeting—is a potentially useful resource. Don't hesitate to call them for information or support, and be willing to reciprocate the favor.

To expand your network, make an effort to serve on task forces, committees, and projects which give you a broad company perspective. Such assignments increase your "visibility" (others' awareness of you and your performance) and your "exposure" (your awareness of what's going on elsewhere). Experience shows that individuals who identify organization problems or opportunities are often put on task forces which address those issues. Thus participants have literally designed their own opportunities.

Peers and Subordinates

Your peers support your objectives (and vice versa) when you develop collaborative working relationships. The nature of these relationships depends on the structure of the work, the personalities involved, and how the organization handles rewards and recognition.

The obvious value of solid peer relationships is the ability to "borrow" information, knowledge, and skills. Ideally, you can work out informal, mutually advantageous agreements in which each person concentrates on what he or she does best and work is shared to minimize peak workloads on individuals.

Getting the most from subordinates means willingness to delegate. Too often engineers and managers keep the most challenging technical tasks for themselves. As a result, supervisory tasks get insufficient attention, subordinates don't grow, and overall unit effectiveness declines. Delegate the tasks that will free you to concentrate on your supervisory responsibilities.

If you want advancement in the company, delegation is a must. Promotions or lateral moves may be denied if there is no one prepared to step into your job. Enhance your own mobility by preparing subordinates to replace you.

Professional Organizations and Societies

Your personal contacts are a key job mobility resource. You expand your contacts, as well as build new knowledge, by active involvement in scientific, technical, or managerial organizations.

Most such organizations have regular publications and conferences; some have local chapters. An excellent way to broaden your career options is to join organizations outside your own field and industry. For example, biologists in the medical industry could, by attending a food industry research conference, discover intriguing new possibilities to apply their talents to converting industrial feedstocks into fructose or developing crop-resistant herbicides.

Browse through *National Trade & Professional Associations of the United States* to identify interesting organizations. This publication lists several thousand by topic, and describes over 100 associations each under the headings of management, engineering, and research.

Community organizations such as Kiwanis and Rotary are good for making local contacts in a variety of fields and are excellent vehicles for sharpening your leadership skills.

Your "Miscellaneous" Contact Network

Over time, you accumulate a large number of colleagues, friends, college classmates, customers, and so forth. Add to this your local network—the family doctor, grocery checker, banker, neighbors, and others—and you have a formidable information resource.

Joe Girard makes an interesting point in his book *How to Sell Anything to Anybody* (Warner Books, 1980). He notes that if you were to list all the people you know, you could probably identify 250 or more people. If you know 250 people, and they each know 250 who in turn know 250, you have a third-generation resource network of fifteen million people, invaluable for job/career changes. (Chapter 12 describes how to tap this network.)

Your Experience and Skills

Your cumulative base of experience, knowledge, and skills is your most valuable career resource. These skills are the basis of current performance and career mobility. I won't say much more about this here; it's important enough to deserve a full chapter later. (See Chapter 7.) Use these career resources on a continuing basis and work to expand them.

SUMMARY OF KEY POINTS

- Review the terms of your psychological contracts at least annually. Remember, the terms of the contract are changing even if you don't acknowledge these changes. Any contract can be redefined.
- You work for yourself. Use sound principles to manage your career enterprise for success.
- Managing your career enterprise requires that you clarify the services you sell, advertise to create customers for these services, develop a stream of new products, maintain vital records, plan for the future, maintain good human relations, and deliver quality products to meet your financial needs.
- Continually develop and use your career resources so they are available when you need them.

QUESTIONS TO CONSIDER

1   Can you define your psychological or unwritten contract with work? Try to specify what you expect to give and what you expect to get.

2   Are you comfortable with this contract? What aspects would you like to change?

3   Which departments of "You, Inc." could benefit from further development? What specific improvements would you like to make?

4   Is your supervisor aware of and supporting your career objectives? What steps might you take to gain more support?

5   Who are your mentors or potential mentors?

RECOMMENDED READINGS

Scheele, Adele. *Skills for Success*. New York: Ballantine Books, Inc., 1979. Valuable guidelines for becoming more active in your job, building connections, taking creative risks. Includes interviews with notables in a variety of career fields.

Woods, David Lee. *My Job, My Boss, and Me: Gaining Control of Your Life*. Belmont, California: Lifetime Learning Publications, 1981. Practical advice and psychological insights for those experiencing on-the-job difficulties and frustrations.

Gabarro, John J. and Kotter, John P. "Managing Your Boss." *Harvard Business Review*, January-February, 1980. Describes how to develop a compatible working relationship, clarify mutual expectations, and maintain good information flow.

Schmidt, Terry D. *Winning in the 80's: Strategies for Career and Life Success.* New York: AMACOM, 1983. Written for a general non-technical audience. Combines guidance with self-assessment projects.

CHAPTER 4

Understanding Your Organization's Personality

The employer generally gets the employees he deserves.

Sir Walter Gilbey

OVERVIEW

You bring a unique personality and style to work each day. But the organization also has a distinct personality. Your daily satisfaction and future prospects in that organization are based on whether these personalities harmonize or clash. You can function more effectively in your current organization and make better future choices if you understand:

- How to determine the organization's informal "game rules"
- The influence of organization culture on the work atmosphere
- The historical and occupational group factors that affect organization personality
- How to observe company political characteristics which make an organization a good place to work
- Indicators of enlightened human resource policies

ORGANIZATION "CULTURE" AFFECTS YOUR WORK SATISFACTION

The better you and your job "fit," the happier you will be. Square pegs don't like round holes; round pegs suffer in triangular holes. Your "peg" is shaped by personal values, key skills, and career goals; the shape of the "hole" is determined by the skill requirements of the job plus subtle factors which establish an "organization personality."

Every organization you have worked for has a distinct personality and unwritten rules about how people should behave individually and with others. These informal rules determine how things *really* get done, as opposed to the official version of organization structure, policies, and procedures.

These unwritten rules, sometimes called organization "culture," exert an invisible, yet powerful, influence over your day-to-day behavior, acceptance by others, job satisfaction, and career progress. Culture is the key to how well you and your organization fit. Failing to understand and work within the accepted culture can be dangerous. Here's an example of cultural mismatch: Bill Johnson was a top-flight systems analyst who prided himself on his competitive spirit and technical competence. Johnson took great pride in outperforming his colleagues and was rated a superior performer.

When Johnson changed companies, he brought with him his technical savvy and competitiveness. Johnson plunged into the new job with enthusiasm and high expectations. Six months later he was fired.

Johnson fell victim to organization culture. He failed to understand that while the technical work was similar, the unwritten behavior codes were not. His first company appraised individual performance in a zero-sum fashion. This created an atmosphere in which individuals were reluctant to assist their peers.

MR. TWEEDY   by Ned Riddle

"Brad knows that yawning is contagious. He gives Tweedy a hard time when he's having a meeting with the boss."

But the new company appraised group as well as individual performance. Success there required a collaborative rather than a competitive style. Johnson was viewed as interested only in his own work, and unwilling to contribute to other people's and the work unit's objectives. He failed by using precisely the same characteristics which had made him successful in his first company!

Personal adjustments to the organization's culture may be necessary, especially to gain admission to the "inner circle." You can't rapidly change culture—it evolves slowly. But the unwritten laws of organization culture are critical factors in large and small decisions which affect your work.

By better understanding the rules of the game, you can better play the game—or decide the game is not for you.

HOW ORGANIZATION CULTURE IS FORMED

Organization culture is affected by its stage in the industry life cycle. (See Chapter 2.) The business demands of each stage influence the organization's character, as the following example shows.

"It's amazing how this company has changed in the past 10 years. There were only 100 people and one production plant when I joined. It was an exciting place. We were the best damn organization in an exciting new technology. We made a decision one day and put it into effect the next—we ran lean and mean. We grew like crazy—sales doubled annually for several years.

"It's a lot different now. We have over 3,000 people and 12 plants. Growth has almost stopped; we're lucky to increase sales by 5% a year now. It's not the same company at all. It takes forever to make a decision. The spirit we had in the early days is gone. I hate to admit it, but we've become a bureaucracy. Managers used to take risks; now they cover their behinds. The hotshots we had in the early days have all left—I guess they were entrepreneurs at heart. I've been able to adjust okay. In fact, I enjoy not having to work so hard—it gives me more time for other things."

Organization characteristics vary with stage of growth, as shown in Figure 4–1. If you thrive in an informal, risk-taking environment, you will be most satisfied in emerging and growth industries.

On the other hand, if you are most effective in a more stable environment, look to the mature industries. Avoid industries in the declining stages if your career has many years ahead. Growth stage also affects salary and promotion patterns, as discussed in the next chapter.

In smaller and newer companies, culture is shaped by the philosophy of the organization's founders and key members. A company founded by inventors, for example, would encourage more R & D lab innovation than one whose success is due to aggressive marketing.

Figure 4–1 Management characteristics vary with organization growth stage

Characteristic	Growth stage			
	Emergence	Growth	Maturity	Decline
Management style	Strong, centralized entrepreneurship	High delegation and freedom	High delegation, fixed goals	Low delegation and freedom
Planning and control systems	Highly informal	General goals	Clear goals	Tight quantitative control
Motivation and risk-taking	Highly venturesome, risk-taking	Business growth, measured risk	Conservative, risk averse	Very limited risk-taking
Decision making	Few formal goals, limited information	General goals, more information	Clear goals, information-based decisions	Rigid goals, information for control purposes

Source: Management Analysis Center, Inc., Cambridge, Mass. Used by permission.

Organization culture also reflects the personalities, values, and thinking styles of the company's dominant occupational groups. Thus, a well-run organization heavily populated by engineers would encourage an internal culture conducive to good technical performance. In a firm where the engineers have less influence than, say, the sales force, this might not be true.

Wherever you work, learn about that culture. Ask a long-time employee to give you an "oral history" of the company. Assess your degree of compatability with the company culture. If you discover a gap, you have three choices: change yourself, change the situation, or change jobs. If you opt for job-change, be sure to test the new culture before you switch. Job-changers who don't may discover that despite a better salary or title, the new job doesn't fit.

THE PERSONALITY STYLES IN AN ORGANIZATION

Dr. John Holland, a pioneer in the field of vocational choice, believes that individuals select work environments consistent with their key personal attributes. His theory is supported by empirical studies demonstrating that certain personality types choose certain occupations.

Holland's six personality types, and occupations typical of each, are:

1 *Realistic* People with strong mechanical or athletic abilities who prefer to work with machines, tools, objects, plants, or animals. They often enjoy working outdoors. Civil, industrial, and mechanical engineers fit this group, as do architects, urban planners, and medical technologists.

2 *Investigative* Problem solvers, they like to observe, investigate, analyze, and evaluate physical, biological, and cultural phenomena and are strong in science and mathematics. Chemical and electrical engineers, physicists, chemists, mathematicians, oceanographers, geologists, materials scientists, and psychologists fit the investigative category.

3 *Artists* Innovative and intuitive, they seek to create through self-expression. Writers, drafters, musicians, actors, painters, and sculptors are found in this group; few technical disciplines are represented.

4 *Social* Like to work with people—informing, training, developing, curing, and enlightening them. They develop interpersonal, rather than intellectual or technical skills. Again, few technologists but many political scientists, medical assistants, psychologists, teachers, clerics, sociologists, and health workers.

5 *Enterprising* Like to manage, influence, lead or persuade people to attain organization goals. They have strong leadership and speaking skills, but are generally deficient in scientific competencies. Engineering managers, marketers, general managers, administrators, business educators, lawyers, and politicians fit the enterprising category.

6 *Conventional* Collect and analyze data and carry out structured, systematic, analytic activities. Strong in attention to detail, they value business and economic achievement. Computer scientists and systems analysts fit this category, as do statisticians, accountants, financial analysts, and administrative assistants.

Holland collected extensive data on personality styles and vocational choice. He noted that most personalities overlap rather than fit into only one category. As you scan these six categories, you are apt to find two or three which describe you. Holland found some interesting correlations among variables of different personality types.

The hexagon shown in Figure 4–2 is based on this analysis. Adjacent categories in the hexagon (e.g., realistic-investigative, conventional-enterprising) are highly similar, while diagonal categories (e.g., investigative-enterprising, social-realistic) are highly dissimilar.

Figure 4–2 Holland's hexagonal model of the
relationships among occupational personality types

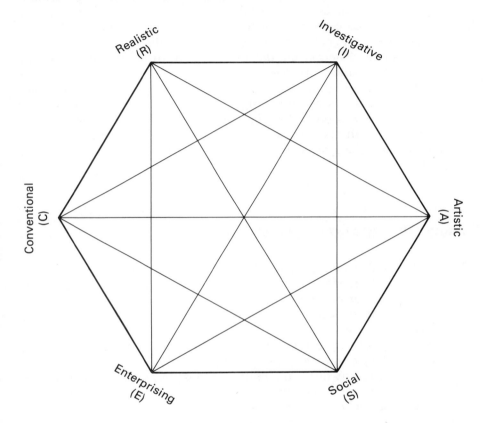

If your dominant personality types fall into one or more *adjacent* categories, you are more likely to be most content with your career choices. You will discover your career "niche" with greater ease and experience fewer confcts in making choices. If your personality characteristics fall in diagonal categories, you are likely to have internal conflicts about your choices. You are apt to have greater difficulty discovering what fits you best, because you could function in many different roles and any choice you make will satisfy only part of your needs.

This model helps explain why superb scientists and engineers are often unhappy as managers, and why the best managers are often mediocre technologists. The management-oriented person, Holland suggests, is verbally skilled, ambitious, aggressive, domineering, and aspires to status and power—the enterprising category. The next most effective manager would come from the social (strong in interpersonal and human relations skills) or conventional (conforming and rule-oriented) categories. But the best scientists and engineers fit the realistic and investigative category. Their strengths are in logical thinking, investigation, analysis, and technical problem solving.

Better understanding your own personality type and that of others in the organization is a key to improved career decision making. The questions are: Are you comfortable with the people? Are there enough different personality types to keep things interesting and enough people similar to you so you are comfortable? That's what personality/culture fit is all about.

ANALYZE ORGANIZATION POLITICS

Many technical people consider "organization politics" a dirty word, something that interferes with doing a good technical job. At its worst, politics is a vicious process which rewards the wrong kind of individual performance. At its best, it is a useful process which forces compromise and acceptable decisions.

Unfortunately, ignoring company politics won't make it go away; it's part of every organization's culture.

You should be aware of the political rules your organization culture allows and encourages. Organization politics boils down to power relationships and pecking orders among individuals, offices, and departments. A good way to understand these relationships in your organization is to study the decision-making flow and watch for the visible "perks" of power. Questions like the following are useful:

How are important decisions made? Who is consulted before the fact? Who is informed after the fact?

Who can veto a decision? How, when, and by whom are vetoes overturned?

Who must sign off on key documents, proposals, and correspondence? In what order?

Whose subordinates tend to get promoted frequently?

Whose offices are nearest to top management?

Who spends time with top brass at work and socially?

Who sits on important committees and task forces (individuals, offices, departments)? Who chairs these groups?

Who gained and who lost power, budget, or staff in the most recent reorganization?

Who has the best furnished office and reserved parking spaces?

Who gets mentioned most often in company periodicals?

The questions cited above will help you identify political patterns; the question of how to respond must be answered in the context of your personal values.

Personally, I'll play the game to a limited extent. I try to get along with most people, and avoid making enemies. I don't hesitate to make my accomplishments visible to those who wield the clout and I won't let others take advantage of me. But I'm not willing to back-stab my colleagues or brown-nose the higher-ups. For a practical reason, I avoid organizations where success depends more on who you know than how you perform. Such organizations drive out their best people, and can't be successful in the long-run.

If you enjoy organization politics, seek environments where political gamesmanship skills are a plus. If you don't, look for organizations where those factors are less important. But if you avoid political involvement, you must still be willing to speak up when a colleague claims the credit for your work, confront situations where you are treated unfairly, initiate a job renegotiation discussion, and neutralize the tactics of those who attempt to manipulate you. The interpersonal factors in Chapter 14 are both offensive and defensive political tools.

Actual power doesn't just correspond with position or title. Many individuals who don't appear on the organization chart exert considerable influence through their specialized knowledge, personal charisma, or informal network. Doing your job exceptionally well is the best way to increase your power base.

ORGANIZATION CHARACTERISTICS TO LOOK FOR

Find or help to build an organization culture where you can thrive. Here are some organization characteristics I believe are important for personal satisfaction and career growth:

- Recognition of your individual contributions and status as a professional
- Opportunities for significant achievement or contribution
- Company reputation as a capable, professional organization
- Good use of your talents, with a minimum of routine and burdensome paperwork

- Monetary and psychic rewards for superior performance
- Interesting colleagues and a variety of ideas and viewpoints
- Organizational support for personal growth and new skills development
- Supervisors and management who are willing to listen to new ideas
- Open decision making, where important decisions are arrived at through open discussion and are based on merit rather than formal power

For an interesting self-insight, list organization environment factors you consider important. Add a ranking or priority scheme to the individual factors. This list is useful for appraising how you feel about your current work, and invaluable as criteria for job-change planning.

Don't overlook the opportunity to enhance these factors where you work. Your actions and attitudes help shape the environment. An individual whose job environment lacked exposure to new developments suggested the office adopt a weekly "brown-bag" lunch session, inviting persons from other parts of the company to explain the latest developments in their field. The idea spread company-wide, and provided many benefits—exposure to cutting-edge ideas, awareness of new business development areas, practice in making presentations, and so forth. Synergy in action through one simple idea.

REVIEW HUMAN RESOURCE DEVELOPMENT STRATEGY

It is no coincidence that the most successful companies—the IBMs, the General Electrics, the Hewlett Packards, and the like—also have first-rate strategies for developing and using human resources.

For decades, organizations have made new plant and equipment investment decisions by calculating the productivity benefits and economic rate of return. Only recently have organizations realized that careful investments in human resources also pay productivity dividends.

A good indicator of a progressive organization is its policy of investing in the continuing growth and development of its employees. Does the company provide on-the-job technical and managerial training? Pay for seminars and sabbaticals? Encourage attending conferences and professional meetings? Can you learn on company time or are you expected to use your own?

Equally important are policies which ensure that people do not become "stuck" in any job for too long. Does the company encourage employees to set targets for their next job, and provide a variety of development experiences? Look for situations where professional employees are rotated for job variety and growth.

You can't expect smaller companies to offer the same development policies as larger companies. The specific job assignment is the critical factor in your growth. If the job provides challenge, variety, and growth, formal learning experiences are less important.

SUMMARY OF KEY POINTS

- Your work atmosphere is affected by informal organization culture climate factors. Identify these unwritten "rules of the game" to play it better or decide the game is not for you.
- Your own attitudes and efforts can change the game rules in your work unit. Work to create a healthy, stimulating, productive climate—you don't need formal authority or position to do this.
- Your personality type provides a clue to understanding organization "fit" and explains why some people are unsuited to particular jobs.
- Clarify the organization characteristics important for your satisfaction, growth, and effectiveness. Review your current organization against these characteristics, and use them as criteria for future choices.
- One of the best ways to improve your power base is to do your job exceptionally well.

QUESTIONS TO CONSIDER

1   How would you describe your organization's personality?
2   What informal rules determine how things really get done in your organization?
3   What type of behavior is necessary to be effective here?
4   Do any aspects of these informal rules conflict with your own values? How serious are these conflicts?
5   What is the commonly accepted concept of how people get ahead in the organization? What does "getting ahead" mean in the company? What does it mean to you? What accounts for the success of the most successful people in the organization?
6   Do you have sufficient power to accomplish your objectives? Whose influence or cooperation do you need, and how can you get it?

RECOMMENDED READINGS

Kolb, David A., Rubin, Irwin M., and McIntyre, James M. *Organizational Psychology: A Book of Readings.* Englewood Cliffs, New Jersey: Prentice-Hall, Inc., 1974. A collection of interesting articles on organizational climate, organizational leadership, organizational decision making, and learning theory. The articles are written by leading theorists and have a heavy research orientation.

Zaleznik, Abraham. "Power and Politics in Organizational Life." *Harvard Business Review*, May-June, 1970. A thoughtful analysis of the use and misuse of position power. Good discussion of how power can be used intelligently to enhance problem solving.

Kennedy, Marilyn Moats. *Office Politics: Seizing Power, Wielding Clout.* New York: Warner Books, Inc., 1980. A "how-to" course in office politics, with good discussion of how to recruit allies throughout the company, and how to prevent power tactics from being used against you.

CHAPTER 5

Learning the Compensation/ Promotion Game Rules

Money is a good servant but a bad master.
H. G. Bohn

OVERVIEW

How does the salary game work? What factors determine how much money an individual receives? How can you enhance promotion prospects? This chapter answers these questions by discussing:

- The fundamental "law" of personal compensation
- Industry, company, and other variables which affect salary
- Industries which maximize income and promotion prospects
- How to influence the reward system
- Tactics for getting a raise

THE FUNDAMENTAL LAW OF PERSONAL COMPENSATION

Surveys of U.S. workers during the 60's and 70's asked people to rank their most important work motivations. The most frequent responses were chance to contribute, personal satisfaction, challenge, opportunity to advance, and similar psychic rewards. Income invariably ranked fifth or sixth. If that same survey were taken today, I'm sure that the importance of income would rise a few notches.

Many professionals who used to enjoy real income increases today find too much month left over at the end of the paycheck. The uncomfortable pinch of double-digit inflation, bracket creep, and economic uncertainty has rekindled the importance of compensation in career strategy.

Seldom are the howls of anguish so loud as when the company announces its salary, raise, and bonus decisions. "But I did everything in my job description and everything I was told. How can they say my performance was only adequate?" And another: "Parker didn't do any more than I did. How come his raise was bigger?" Still another: "How do they expect me to afford my new boat with this measly increase?"

What Value Do You Add?

Each organization has its own inscrutable methods for determining financial rewards. These methods attempt to convert subjective observations into objective formulae, which divide the employees into categories for increases above, below, or equal to a predetermined "control" rate. The methods are never totally fair, but it's the only game in town.

I'd like to suggest an underlying principle that determines individual compensation. It goes something like this:

> You are compensated according to the *value* you add to your organization.

It is clear how the principle applies to athletes and entertainers. Baseball player Dave Winfield landed a million-dollar-plus-annual contract with the New York Yankees because, in the judgment of Yankee management, his performance at the plate and in the outfield would attract enough extra fans to justify his salary. Bob Hope earns $25,000 a night at Las Vegas clubs because he draws large audiences who will spend in excess of this.

This same principle applies to scientists and engineers. If you are a technical genius whose fertile mind constantly cranks out new invention, and if the company exploits these products to reap high sales and profits, you are entitled to—and will get—your share of the largesse. If you are a marketing representative in a company whose reps average a million dollars worth of sales a year, and you find a new customer whose first order is five million dollars, rest assured your compensation will far exceed that of the others.

These examples are apparent because the value added is easy to quantify. Persons in sales or other positions which directly affect sales or profit tend to be paid better, in part because the value added is easier to measure.

Prove Your Value

This basic principle, however, applies to virtually every job and the value can be economic or other. If the value *you* add—as defined by the organization—is less easily measured, it's your responsibility to demonstrate that value. This requires doing well the tasks that are in your job description, plus some that are not. The following example shows how individual initiative can bring both psychic and financial rewards.

> "Our production inefficiencies were obvious to everyone, but nobody did anything except complain. One day I decided to do something about it. This task wasn't assigned to me, but I checked it out with my boss and he gave me some time to come up with some ideas. With a bit of observation and analysis, I devised several recommendations which would save the company time and money.
>
> "I described these in a memo to the division head, with my name on it. I submitted it through my boss, who attached a concurring cover memo and passed it on. I made sure that my name stayed on it, but I wanted boss support so I tried to give him a bit of the credit. Know what happened? The recommendations were implemented and they saved over $100,000 the first year. When the annual performance review came, I reminded the powers that be of my work. It paid off handsomely, in a bonus and choice assignments."

Performing only basic job requirements and little more is enough to keep your job, but not enough for extraordinary rewards. The bulge of the bell-shaped curve is full of people who do what the job requires, but not much more. The juicy rewards are reserved for those who contribute beyond what the job minimally requires.

The best way to increase your compensation is to increase the value you add to the organization. Regardless of where you sit and what you do, there are ways to do so. This does not necessarily mean working *harder*, but working *smarter* to reach organization objectives and enhance productivity. It means showing some good reasons why you deserve more money and bringing these reasons to the attention of key decision makers. I'll discuss how to influence the reward process shortly, but first let's examine some other factors which affect salary.

FACTORS WHICH DETERMINE SALARY POTENTIAL

Compensation patterns are also influenced by several industry, company, and geographic variables. Here are some of the most important.

Industry/Company Profitability

Industry and company profit levels determine what they can afford to pay. *Forbes* magazine calculated the median industry profit in the past five years at 4.5%, with some industries far above the average (e.g., drug profits are 8.5%) and others far below (autos and trucks are 2.2%). Industries with higher returns can pay higher salaries for the same jobs. Within an industry, compensation tends to be higher in the more profitable firms.

Profit and other financial performance data are regularly published by *Forbes, Business Week, Dunn's Review, Fortune, Wall Street Journal*, and other major business publications. Appendix III summarizes financial performance by major industry groups.

The Organization's Human Resource Strategy/ Policy

Major corporations determine salary ranges using survey data about other companies in their industry. A company wishing to be seen as a leader sets its median salary a specific percentage ahead of other similar firms. A company in a unionized area may well set its salaries higher to act as a disincentive for unionization.

Importance of the Function to Organization Objectives

The employee's role in the success of organization strategy also affects compensation. A mining engineer teaching at a state university, for example, will earn a salary set by academic pay rates. The same engineer locating new mineral deposits for an exploration firm would command much more, as his efforts could add substantially to sales and profits.

Skills Supply and Demand Patterns

Business cycles, public policy, and technological advance change the demand for technical talent. Skills in short supply command higher salaries; those in surplus drive down compensation for that field. But labor market flows tend to balance supply and demand into equilibrium as individuals change industries, regions, and jobs. If you are part of, or able to move into a high demand and low supply category, your income will reflect the scarcity economics. (See Appendix II for supply and demand projections in 43 major engineering and scientific specialties.)

Regional Desirability and Economics

Compensation levels reflect the desirability of certain parts of the country and economics of the local area. For example, the San Francisco Bay Area is both a high-pay and a high-cost area. By contrast, the new technological centers in Georgia, Colorado, and Florida are both lower-cost and lower-pay regions. But there is a twist: people in a desirable area may be reluctant to move and when local labor supply exceeds demand, compensation patterns are depressed. Seattle experienced this during the aerospace layoffs of the early 70's.

Higher compensation is required to lure specialists to jobs in undesirable living areas. The attractive salaries offered to communications engineers in Saudi Arabia, for example, reflect the meager living conditions there.

Life Cycle and Growth Rate

New and expanding industries generally offer better salaries and salary growth; mature or declining industries pay less. Rapidly-growing companies must recruit externally and offer higher salaries to attract new personnel. The fastest-growing companies in higher-growth industries usually provide the best compensation.

FACTORS WHICH DETERMINE PROMOTION POTENTIAL

Your own promotion prospects depend on many factors—the frequency of suitable openings, the number of different positions for which you qualify, the nature of the competition, company policy on internal promotion versus outside hiring, whether and how upcoming openings are announced, mechanics of the selection process, and so forth.

Charles Guy Moore, a former Senior Economic Analyst at Exxon, has identified five strategic characteristics of a job that make it a likely or unlikely springboard to rapid career advancement. Dr. Moore's five-year study of 200 successful professionals indicates that your firm's growth rate is the single most important determinant of rapid career advancement.

Promotions are most common and come soonest in growing organizations: growth creates additional positions, and you also become a relatively more experienced employee the faster your company grows. In stable organizations, promotions is limited to positions created by turnover or retirement.

The promotion advantages of working for a growing company are shown in Figure 5–1, taken from Dr. Moore's book, *The Career Game*. To interpret this table, look down the far left column until you come to the number of years you've worked for the company. Then go across to the appropriate column that represents your firm's growth rate. The intersection of these shows the percentage of employees with more experience than you have. For example, if you had worked 8 years for a company that was growing at the rate of 16% annually, only 31% of its employees would have more experience than you. If you had worked the same 8 years for a company that was only growing at 4%, however, 73% of its employees would still have more experience than you.

Figure 5–1 How growth affects your status as an experienced employee

Your experience in years	Percentage of your firm's employees who have more experience than you, assuming your firm grows at a compound rate of growth of:						
	4%	8%	12%	16%	20%	24%	28%
0	100	100	100	100	100	100	100
1	96	93	89	86	83	81	78
2	92	86	80	74	69	65	61
3	89	79	71	64	58	52	48
4	85	74	64	55	48	42	37
5	82	68	57	48	40	34	29
6	79	63	51	41	33	28	23
7	76	58	45	35	28	22	18
8	73	54	40	31	23	18	14
9	70	50	36	26	19	14	11
10	68	46	32	23	16	12	8
11	65	43	29	20	13	9	7
12	62	40	26	17	11	8	5
13	60	37	23	15	9	6	4
14	58	34	20	13	8	5	3
15	56	32	18	11	6	4	2
16	53	29	16	9	5	3	2
17	51	27	15	8	5	3	2
18	49	25	13	7	4	2	1
19	47	23	12	6	3	2	1
20	46	21	10	5	3	1	1
25	38	15	6	2	1	0	0
30	31	10	3	1	0	0	0

Reprinted with permission from *The Career Game* by Charles Guy Moore. The revised edition of this 275-page book is available at $9.95 from the National Institute of Career Planning, 521 Fifth Avenue, New York, NY 10017.

While organizations attempt to promote the most qualified individuals to open positions, the selection process has a subjective dimension as well. In addition to performance, promotion choices may be influenced by:

- Attending certain schools and having the "right" credentials
- Involvement in high-visibility projects or assignments
- Working in the "right" departments or offices
- Being connected in certain networks
- Working for a boss with "clout"
- Being on the winning project at the right time

To better understand the importance of these factors, consider:

- How are the promotion decisions really made?
- Who are the people who make such decisions?
- What type of people seem to get ahead in this organization?
- What do they do to get ahead?
- Which sequences of positions or experiences provide faster progress than others?
- Are you blocked because you work for the wrong boss, office, or department?

Plotting Your Promotion Strategy

If promotion is important to you, you must become known to those who make such decisions. One enterprising individual left his office each day the same time the key decision-makers left and used the walking time to the parking lot to introduce himself and describe his work. As a result of one such "chance" encounter, he was assigned to a high-visibility task force which led to a subsequent promotion.

To plot your promotion strategy, first study the nature, number, and departmental distribution of jobs in your company. Visit the personnel department for information on job families, skills requirements, typical progression paths, and future work force projections. While every organization has many different jobs, they cluster into job families with similar skill/knowledge/experience requirements. To increase your promotion probability, identify categories of similar jobs rather than selecting just one or two specific positions.

Pay special attention to how the job mix will change in the future. Organizations project their personnel needs by the number and types of skills needed three to five years in the future. Concentrate your learning and development efforts on company growth areas. This strategy makes sense even if you're not sure you'll stay with the company that long. If your company's skill needs are shifting, other firms in the industry are experiencing similar changes. Thus you increase both internal and external opportunities by concentrating on growth areas.

If one of your job targets is your boss's job, your strategy is to get him or her promoted and become the heir apparent. Promoted employees often name their own replacement, and by contributing to your boss's success you increase your own chance of being selected. As a crucial subordinate, you may move as your boss progresses. Many have enjoyed rapid career progress by hitching themselves to a "comet."

The principle I discussed in Chapter 3—working to achieve organization objectives your boss is responsible for—enhances your prospects of becoming the replacement. Try to get your boss to delegate additional tasks to you.

If you want to move up, become replaceable. If there is no one ready to take over your job, begin training someone. Promotion may be denied if you are the only one capable of doing your work. Nature abhors a vacuum; organizations do too.

Don't base your own view of career success on receiving automatic and regular promotion. The hard realities of the 80's—slower economic growth and a larger, better-educated work force—mean that many people capable of holding higher-level jobs will not get the chance to rise as far as they would like. Promotion is one possible result of doing well what you enjoy doing, but not the most important factor in experiencing career success.

USE PERFORMANCE APPRAISAL TO YOUR ADVANTAGE

The performance differences among a group of people doing similar work may be slight, yet certain individuals get the nod when it comes to promotions, pay raises, and choice assignments. By properly using

© 1981 Field Enterprises, Inc. Courtesy of Field Newspaper Syndicate

the performance appraisal process, you can influence these decisions in your favor.

Understand What the Appraisal Measures

The formal reward process in most organizations centers around an annual or semi-annual performance appraisal. Effective appraisal systems measure accomplishment of specific objectives jointly set by and agreed to by each employee. Less effective systems measure personality characteristics (such as cooperation, attitude, initiative, decision making) rather than specific accomplishments.

In either case, you should clearly understand the basis for your appraisal. Keep a journal of your work accomplishments, and provide those who evaluate you with data to simplify their task. The following example shows how this can pay off.

> "In my work unit there are eight people with identical job descriptions. I know that my boss hates to fill out that blank appraisal form each year. I can see why—after a year it's hard to remember the accomplishments of each person. This year I made it easier for him. Three weeks before the appraisal, I wrote a memo summarizing my accomplishments for the year and gave it to my boss.
>
> "When I reviewed my completed appraisal, the form was full of specific accomplishments—the same ones I had submitted. Other people in the unit do similar work, but their appraisals didn't reflect specific achievements. When the appraisals went to the next level, mine stood out and I was rated high. Actually, I didn't achieve much more than the next guy. I just reminded the boss what I *did* do."

View Appraisal as a Three-Step Process

Make the most of the performance appraisal process. Regard it as a tool for you, not a threat. View performance appraisal as a three-step process, rather than a one-shot event.

Agree on Objectives

Develop a shared understanding with your boss of how you will be measured at the beginning of the appraisal period. Make sure you agree on key tasks and their means of measurement. It does no good to perform your tasks superbly if, in your boss's eyes, those are the wrong tasks. The beginning of the cycle is an opportune time to revise the scope of your work to include opportunities of interest to you. You may have to initiate this discussion, but you are doing both of you a favor.

Reduce the subjectivity of subsequent appraisals by coming up with specific accomplishment measures. Turn generalized objectives into concrete measures using quantity, quality, and time indicators. For instance, if your job includes writing technical research proposals, turn this into an objectively verifiable measure, such as "write not less than two technical proposals quarterly (a quantitative and time measure) of which 50% are approved by higher management (a quality measure)."

Use this discussion to reach agreement on how performance will be measured. Discuss what constitutes satisfactory performance, and what constitutes superior performance. If 50% acceptance is satisfactory, what percentage would warrant a superior rating in this job dimension? Such clarification reduces the subjectivity which frequently sours the appraisal process.

Interim Progress Reviews

These are the second part of the process. Over time, priorities change and you get new assignments. If so, your job standards should be revised. Review progress thus far, and modify the objectives as needed. Interim reviews should be held at least every six months, and preferably quarterly. Again, you may need to take the initiative in this discussion.

The Formal Appraisal

This is the third part. If you started with clear and agreed standards, held interim reviews, and summarized your accomplishments for the year in a memo, the formal appraisal won't contain any surprises.

This three-part process works because so few people actually use it. It lets you concentrate on what's important (in the eyes of the boss), makes you stand out as one concerned about ongoing improvement, encourages a collaborative working relationship, builds trust, and simplifies the boss's job in appraising you. The process benefits both parties.

HOW TO GET A RAISE

The best time to bring up the issue of a raise is when you have been assigned an additional workload or expansion of responsibilities. Approach it from the standpoint of "Here is how I plan to carry out these responsibilities. . . . By the way, how does my increased responsibility translate into compensation?"

Supervisors get uncomfortable discussing such things and usually cite "company policy" reasons why they cannot comply. What this really means is that your expectations aren't in their current budget and they need approval at a higher level. Your objective is to demonstrate why you should constitute an exception to normal "policy."

Do some homework first. Through your informal network, find out what other companies pay for similar work. Also check to see how company sales and profits are going. Don't raise the issue if the company

is barely surviving—pick a time when the company can afford to meet your expectations.

Keep your cool and negotiate effectively. Exploding in anger or breaking down into tears doesn't help. Mentally rehearse this discussion, and prepare counter arguments to all possible objections. The communications skills discussed in Chapter 14 apply to this discussion.

Don't use the "give me more money or I'll quit" ultimatum unless you are prepared to follow through. It doesn't hurt to hint that you have been approached by another firm, and though you don't want to change jobs, the other offer is so financially attractive that you can hardly afford to stay here . . . what can the company do about compensation?

If you don't get the money you want, avoid the sudden impulse to jump companies for a few bucks more. You have made an important investment in your company, and there are inherent advantages in staying. Raise refusals may not be permanent. In fact, the salary differential required to make a job change financially advantageous is significant in today's economic climate, especially if it involves relocation.

This chapter discussed promotions and the financial dimension of compensation. But beyond an income sufficient for a reasonable standard of living, I much prefer additional compensation increments to be psychic rather than financial. As Robert Southey wrote in *The Doctor*, "As for money, enough is enough, no man can enjoy more." To me, it's much more important to enjoy what I do and feel good about the people I work with than to earn a little more or to have a fancier title.

SUMMARY OF KEY POINTS

- Your compensation is based on the *value* you bring to the organization. There are some broad concepts affecting value, but in the end it is defined by the organization.
- Doing the best job you possibly can is only the first step. Making sure the decision makers *know* about your accomplishments is the vital second step.

- If you are after frequent promotions and upward mobility, look to the emerging and rapid-growth industries.
- Use the performance appraisal process to develop a shared understanding of job expectations. Reduce the subjectivity of appraisal by making your objectives measurable.
- Keep a journal of work accomplishments. Forward this information to assist your supervisor in reviewing your performance.
- Don't hesitate to ask for a raise, but make sure you have some good reasons why the company should say yes.
- Don't job-hop for only a few dollars more; the "psychic" income factors have more effect on your career satisfaction than modest financial differences.

QUESTIONS TO CONSIDER

1   In what stage of the life cycle is your industry? Is the overall industry growing, declining, or staying the same?

2   How does your company growth rate compare with the industry average? How does the industry/company life cycle affect your career?

3   What are some ways you add value to the company? How could you increase the value you add?

4   What are your personal salary needs and desires? At what level does increasing psychic income become more important than increasing dollar income? Have you already reached that level?

5   In what other kinds of situations (career function, industry, job) could you bring more value to a company?

RECOMMENDED READINGS

Castain, Sherry. *Winning the Salary Game: Salary Negotiation for Women.* New York: John Wiley & Sons, Inc., 1980. Don't be misled by the title—the strategies here apply equally well to men. Tactics for getting more money where you are and negotiating salary when you change jobs.

Van Caspel, Venita. *Money Dynamics for the 1980's.* Reston, Virginia: Reston Publishing Co., Inc., 1980. Practical strategies for coping with inflation, investing, and winning the money game. Thorough, well-written descriptions of stocks, bonds, mutual funds, insurance, and related topics.

American Association of Engineering Societies, Inc., 345 East 47th Street, New York, NY 10017. This federation of 39 engineering societies publishes annual surveys of engineering salaries and other useful reports on technical employment.

PART TWO

MAKING STRATEGIC
CAREER DECISIONS

CHAPTER 6

Developing Your Career Strategy

The future is not some place we are going to, but one we are creating. The paths to it are not found but made, and the activity of making them changes both the maker and the destination.

John Schaar

OVERVIEW

Career strategy provides guidelines for decision making and action. The best strategy is flexible, realistic, and geared to your goals and values. This chapter probes:

- The purpose of career strategy
- Factors to consider in developing your strategy
- The role of experience in refining your strategy
- Importance of career options and mobility
- How to observe patterns and themes in your career to date

CAREER STRATEGY IS FLEXIBLE

Career strategy is a set of flexible themes, rather than rigid rules, to guide your career planning. Strategy provides you with criteria for considering alternatives, evaluating options, making choices, and measuring success. Effective strategy gives you satisfaction from the journey, not just in reaching the destination.

Strategy is not developed all at once; it evolves. It can take considerable time, trials, and errors to find your "niche." Management expert

Peter Drucker comments on the issues involved in clarifying your strategy*

"Here I am 58, and I still don't know what I am going to do when I grow up. . . . the only way to find what you want is to create a job. Nobody worth his salt has ever moved into an existing job.

". . . you know what you don't want to do, but what you do want to do, you don't know. There is no way of finding out but trying . . . one doesn't marry a job. A job is your opportunity to find out—that's all it is.

". . . I think one of the most important things would be to know if you like pressure or if you cannot take it at all. There may be people who can take pressure or leave it alone, but I have never met any of them. I am one who needs pressure. . . . If there is no deadline staring us in the face, we have to invent one. I am sluggish, lethargic, a lizard, until the adrenalin starts pouring. A low metabolism—psychologically.

". . . You have to know whether you belong in a big organization. In a big organization, you don't see results, you are too damn far away from them. The enjoyment is being a part of a big structure. . . . And I think you need to know whether you want to be in daily combat as a dragon-slayer or if you want to think things through, to analyze, prepare. Do you enjoy surmounting the daily crisis, or do you really get your satisfaction out of anticipating and preventing the crisis? These things I believe one does know about oneself at age 20.

"There is one great question I don't think most young people can answer: 'Are you a perceptive or an analytical person?' This is terribly important. Either you start out with an insight and then think the problem through, or you start out with a train of thought and arrive at a conclusion. One really needs to be able to do both, but most people can't. I am totally unanalytical and completely perceptive. I have never in my life understood anything that I have not seen . . .

"No matter what job it is, it ain't final. The first few years are trials. The probability that the first choice you make is right for you is roughly one in a million. If you decide your first choice is the right one, chances are you are just plain lazy . . .

"Contrary to everything that modern psychologists tell you, I am convinced that one can acquire knowledge, one can acquire skills, but one cannot change his personality. Only the Good Lord changes personality—that's His business."

Peter Drucker's comments stress the importance of *learning from your life and work experience.* By doing so, the quality of your strategy improves as you learn more about what kind of person you are, the situations in which you flourish, situations to avoid, and so forth. Such self-insights improve strategy.

*Mary Harrington Hall, "A Conversation with Peter Drucker," *Psychology Today,* March 1968, pp. 21ff. Copyright © 1968 Ziff-Davis Publishing Company. Reprinted by permission.

The purpose of strategy is to *increase your probability* of experiencing career satisfaction, growth, enrichment, and success. A well developed strategy helps you to:

- Build an increasing number of future job and career alternatives
- Reduce the number of desirable alternatives you forfeit
- Take advantage of attractive alternatives
- Make more effective choices and decisions

Think probabilistically. Don't confine your future job thinking to a few narrow choices. The U.S. Labor Department says there are 100 million jobs in 20 thousand categories in 14 million organizations in the U.S.—more than enough to choose from. The most intriguing possibilities may be in situations you haven't seriously considered. Broaden your thinking to increase the probability of lifelong work challenge and satisfaction.

GRIN AND BEAR IT   by Lichty & Wagner

©Field Enterprises, Inc., 1981

"There's a 1 in 5 chance that 2 of 3 of you will be fired if 1 of 2 people are appointed the new chief."

© 1981 Field Enterprises, Inc. Courtesy of Field Newspaper Syndicate.

BUILDING BLOCKS OF STRATEGY

Strategy formulation begins with your career "building blocks." Four such building blocks provide useful criteria for considering various options and making decisions. These are your skills and knowledge base, your personal values and beliefs, your work style and personality, and your geographic preferences.

Your Skills and Knowledge Base

Nothing succeeds like success, according to an old saying. Your best prospects for future work satisfaction and success are in jobs that draw on skills and knowledge you use well and enjoy using.

Anyone who has worked for a few years has accumulated a wealth of skills and knowledge. But people tend to take their most important skills for granted. They may be painfully aware of their weaknesses—these get pointed out—but unaware of their real strengths.

You should recognize your best skills, because these are usually transferable to different work situations. Some jobs in virtually every industry require the skills you enjoy using most. Skill awareness reduces job mismatches and increases your job mobility.

Your Personal Values and Beliefs

Your important values and beliefs can help you choose work supporting your "mission." We all have values and beliefs, though we seldom consider them in work choices. Some jobs provide the fulfillment of promoting beliefs important to you. A case in point:

> "The energy crisis made me realize that we can't run forever on fossil fuel resources. I have become a real advocate of renewable energy resources—wind, solar, geothermal, and so forth. I put this belief into practice. My former job was with a big company, developing exotic metal alloys to increase electrical conductivity. Since then, I joined a small, solar manufacturing outfit and am engaged in photovoltaic cell design. I get real satisfaction from knowing I'm doing something important for the world."

You'll be more satisfied if you find work which enhances your values. Most jobs are neutral in this regard. But at least avoid jobs which clash with your value system.

Your work decisions reflect personal and family values as well. If you strongly value outdoor recreation in uncrowded environments, you'll exclude crowded geographic areas from your options. If you believe your children should not frequently change schools, you'll downplay opportunities requiring relocation. If your spouse is career-oriented, your job decisions will consider the impact on both careers. Clarify these important values and determine how your work choices fulfill or frustrate them.

Your Work Style and Personality

Do you function better in a large, stable company which provides greater job security at the expense of narrower job variety? Or would you be happier in a smaller organization which can't provide the same stability but may involve a wider range of activities? Your personality characteristics and work style preferences influence the size and type of organizations to consider.

Do you thrive in the competitive atmosphere of business? Or does the more scholarly academic environment fit you? Do you have the personal determination and willingness to take risks needed to start your own business? Do you like to function as part of a larger team, or do you prefer being a one-person show? These and similar questions concerning your work styles are important to answer as you consider future possibilities.

Your Geographic Preferences

Geographic stability is usually less important in early career years, when individuals will gladly relocate for better opportunities. But as one builds a family and community roots, willingness to move declines. Geographic issues can become extremely important later due to family considerations of aging parents, stability of children in school, or location of grown children.

Consider your long-term geographic and mobility preferences. Many technical specialties are confined to hubs of technological or industrial activity. Some are narrowly defined—oceanographic engineers will find few work options in landlocked states. Others, like computer programmers, enjoy a wide geographic demand for their skills.

Nearly 60% of the engineering and technical jobs in the U.S. are concentrated in 10 states. Figure 6–1 shows the distribution by state. During the 80's, technical job growth will be greatest in the West and the Sunbelt and slowest in the energy-deficient Northeast.

If you plan to relocate but don't know the opportunities in your preferred area, write to the local or state Chamber of Commerce. Ask for industry directories and information on company relocations or efforts to stimulate technical industry.

To make a geographic move without sacrificing job progress involves special strategic planning, perhaps making an interim tactical job move,

Figure 6–1   Geographic location of scientists and engineers (1980)

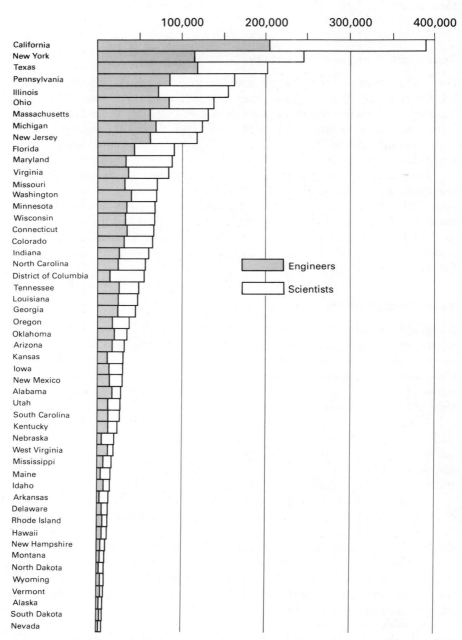

Source: National Science Foundation

as shown in this example. Alan Brock graduated from a West Coast engineering school but accepted a federal government job in Washington, D.C. In the next three or four years, Brock wanted to return to the city where he had gone to school. But he was concerned that his government experience would be harder to "transplant" back to the city he left.

Brock gathered data on rapidly growing industries in his preferred West Coast location. He then identified similar industries in his current region, and directed his job campaign to these. Making this interim strategic move would equip him with suitable industry experience for his eventual return to the West Coast.

YOU ALWAYS HAVE OPTIONS

"Mobility? I don't have much. My background is only average. I've worked here too long to make a change. I don't know what's out there and I don't have any options. I guess I'm trapped."

These comments by a career workshop participant echo the feelings we all may have from time to time—the locked-in feeling. This depressing feeling saps the vitality and satisfaction of far too many people. And, in most cases, it is unnecessary.

We all have far more options than we are aware of. One can engineer changes to the current job or find better opportunities elsewhere in the company. Failing that, one can switch companies, change industries, or move to careers in totally new fields.

Work becomes a prison if you think you're locked in. But what seems like an impossible situation isn't when you recognize that *you always have choices*. Even in the worst situation, you have options. Awareness of these options is a liberating experience.

By becoming aware of these options, you can make intelligent choices concerning whether and what changes to make. You may decide to change or conclude that the best situation is your current job. If you decide to stay, you can do so with a sense of personal control of your career, because you realize that among your multiple options, your present situation is best for you at this time.

Typical Occupational Mobility

If you were to survey your college classmates, you would be amazed at the career flexibility of those with your same background. They will be scattered across the country and around the world, working in diverse industries, performing highly different functions. And they will have made many changes.

If we assume your college peers are typical, their job changes will resemble the pattern identified in a National Science Foundation study of technical occupational mobility during the 1970's. Over a 6-year period, 70% of your peers changed jobs and 26% changed occupations. 35% moved into management, 38% into other scientific/engineering occupations, and 27% entered totally different fields. Figure 6–2 shows the functional mobility by various scientific and engineering disciplines during the study period.

There are as many different patterns of career changes as there are people. The infinite pattern, sequence, and timing of job/function/company/industry/sector/geographic changes you can make defies simple generalization.

This leads to an important conclusion: you can create your own career. There is no best or "right" route; there is only the route you take. The only way to define *best* is in the process of choosing. If you select the best alternatives consistent with self-defined criteria, you are pursuing an optimal strategy, regardless of how a particular choice turns out.

CAREER PATTERNS IN HINDSIGHT

Sometimes the only way to understand your career path is by looking over your shoulder. The important part of career strategy is not your specific job choices but the underlying themes and patterns these choices represent.

MIT Professor Edgar Schein (see the Recommended Readings) studied career changes and identified five general value structures that underpin individual choices. Building from this concept, I have added descriptive labels to each type and oriented the discussion to a technical audience. As you read through the five "types," see which one or two describe you.

- *Technocrats* remain in specific areas of technical or functional interest.
- *Climbers* use their technical skills to climb the organizational/administrative ladder to higher responsibility.
- *Builders* strive to create something of their own, be it a product, service, or company.
- *Searchers* seek to minimize organizational interference and restrictions on their work.

Figure 6–2 Occupational mobility of scientists and engineers

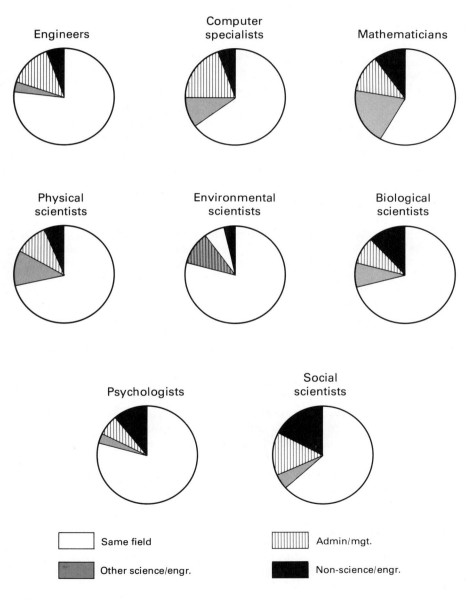

Source: National Science Foundation

• *Stabilizers* value security and are content to remain with their employment organization until retirement.

These types are not mutually exclusive; several may partially describe you. But with more working years, one type usually becomes apparent. By clarifying which pattern(s) fit you, you have useful data for considering future options, making decisions, and measuring satisfaction.

"Technocrats"—Maintaining Technical Excellence

Technocrats' satisfaction comes from the challenge of the work itself. Traditional scientific and engineering values motivate this group—the chance to contribute, interesting colleagues, stimulating work. They seek recognition for their contributions, less through promotions or bonuses than from recognition from other technical people and the broader professional community. As an example, a 36-year-old research biologist with the National Cancer Institute:

> "My work involves state-of-the-art, long-term investigations which eventually save millions of lives. To me, that's exciting—a chance to contribute to something important.
> "My research subspecialty involves recombinant DNA lines. As the field has gotten so complex over the years, I've made several specialization decisions which narrowed my focus. But I've also tracked developments in related research areas. My career goal is to be part of a team that produces a major breakthrough. I can't think of a better place to be for the work I'm in. I'll stay till we totally lick the cancer problem, but I'm not concerned about working myself out of a job soon."

The technocrat's primary loyalty is to the profession and their colleagues, rather than to their current employer. Technocrats tend to shun purely administrative work but will accept project management situations which keep them close to the technical action.

If you share these characteristics, the organization's image is important to you. You are likely to quickly switch employers for more challenging technological opportunities. One career risk to guard against is overspecialization. It is a more severe problem in rapidly changing professions, especially for those who pursue very narrow technical interests. (A later chapter discusses how to guard against this.)

"Climbers"—Up the Organization Ladder

Climbers' key motivation is to advance to the top of the organization. They enjoy the prestige and influence of managerial and administrative jobs. Climbers' possess strong verbal and interpersonal skills which enhance vertical mobility. Climbers maintain technical proficiency only on an "as-needed" basis, and such proficiency decreases as they move further away from front-line technical work. A 32-year-old electronics manufacturing division chief typifies this group:

> "I've probably held more jobs than any one in my graduating class—4 different companies in 3 cities in 12 years. But each one represented a move up, in title and in bucks. Bigger office, thick carpet on the floor, the whole bit. But from here, the pyramid gets narrow, so I'll have to sit a bit or find my fifth company.
>
> "I look at my career as a chess game. I figure out where I want to be several moves ahead and then work that back to the moves I should consider now."

As a climber, you are loyal to the organization as long as there are promotion opportunities. You are willing—even eager—to make tactical moves and jump around. You'll even switch functional areas in your or another organization, if it helps you rise.

Power and influence are climbers' primary motivators. Unlike technocrats, climbers seek respect for their rank and position, rather than their technical acumen.

"Builders"—Creating a Legacy

Builders are inherently creative. They seek situations involving creativity in developing concepts, products, processes, services, or companies with which they closely identify. Consider this 47-year-old mechanical engineer and itinerant tinkerer:

> "I'm a gadget-lover at heart, with more jars of screws, nuts, parts and old machines than my workshop can hold. One day I noticed how much kids watch the television set. This got me thinking, so I built a simple timer which I could program each day to restrict their television watching to an hour or two—after their homework. The neighborhood parents were intrigued by the gadget and before long I went into production operating part time out of my garage. I've applied for a patent and, who knows, maybe I can license the idea to a big company. Or go into full-time production on my own. Meantime, however, I'm not quitting my job."

Like the technocrats, builders also have strong technical acumen. But their predominant drive is to create something unique. This drive may be welcomed by companies which provide funding and administrative support for promising new ventures or products the builder proposes.

If the organization doesn't support their motivations, builders will find a more conducive climate or become entrepreneurs.

"Searchers"—Shucking the Shackles

Searchers find careers in organizations, whether large or small, too restrictive. Searchers may be technically proficient, but find rules, regulations, procedures, and paperwork stifling. They value autonomy and independence above all and are most effective in an environment free of bureaucratic constraints. This 26-year-old technical writer provides a case in point:

> "There's just too much bullshit in big companies these days. I get no kick from pushing papers and putting up with red tape. Life is too short for that. I am my own person, with a distinct work style. I function better when wearing jeans. And I'm a night owl—I work better by dark; during the day my biorhythms seem out of sync. That big company downtown couldn't accept my dress and working habits so it's not surprising I left.
>
> "Two years ago I discovered there's a good market for free-lance technical writers. I can turn a technical phrase or two and I'm doing plenty of little writing jobs with several local firms. I work from my house, mostly at night. Guess what I wear to work?"

Searchers may become "job-shoppers," consultants, or professors, or work for smaller firms less encumbered by procedural restrictions. The decreased cost and increased power of microcomputers creates new opportunities for searchers to establish home-based businesses in computer programming, software development, and the like.

"Stabilizers"—Seeking Security

Stabilizers put career in second, third, or lower priority. Work is necessary to provide income for family and personal pursuits of greater interest. Career progress is less important; security is. They often prefer the current organization—a known entity—rather than risk new environments. They generally maintain performance sufficient to keep the employer content but seldom break new ground. Consider this 53-year-old statistician:

> "I'm looking forward to early retirement in three years and have a reasonable pension coming. I've slayed my organization dragons and had my fill of the competition. I advanced about as far as I wanted to. I stopped climbing many years ago, and that was okay with me.
>
> "My career has been like floating the river in a raft. I've gone through the rapids. Now I float in quieter, less crowded pools downstream. There's enough to keep me interested, and I do my job and keep the company happy. Meanwhile, I'm counting the days until retirement."

While security motivations increase in most people over time, the stabilizer type is not strictly a function of age. Some individuals in their twenties sometimes display these characteristics; others in their sixties are still taking new risks. But stabilizers also need work variety, and should consider periodic lateral changes in the company as a means of new job stimulation with minimum risk.

You may fall into more than one category. There is no one "best" category; all provide satisfaction and success. The important thing is to discern the underlying values in your life to date, and use such knowledge in future decisions.

SUMMARY OF KEY POINTS

- Strategy provides themes and flexible guidelines. Good strategy is molded like *clay*, to give structure and flexibility. It is not rigid like *concrete*.
- A career strategy evolves through a process of sequential decision making. Each decision-making cycle improves your ability to make better future decisions.
- Don't worry if you haven't found your work niche. It can be a lifetime process, but there is no hurry as long as you get satisfaction from the journey.
- Define—and redefine—your career building blocks: motivating skills knowledge, values and beliefs, work styles, and geographic preferences.
- You need never feel trapped in your job: you always have options. But to identify those options and create more requires initiative.
- Your work choices to date reflect your underlying values and work needs. Awareness of these values helps improve future decisions.

QUESTIONS TO CONSIDER

1   What major career decisions are you now considering?
2   What are your key criteria for making these decisions?
3   Does your spouse's work affect your strategy? How?
4   What are your current job options? Brainstorm to make a list of at least ten other jobs which might interest you.
5   Reviewing your career to date, which one or two of the patterns described are similar to yours? Which values or concerns have been at the base of your past decisions?

RECOMMENDED READINGS

Souerwine, Andrew H. *Career Strategies: Planning for Personal Achievement.* New York: AMACOM, 1978. Full of useful information for moving ahead while remaining with one organization. Excellent treatment on dealing with the boss and others on the job.

Greco, Ben. *How to Get the Job That's Right for You: A Career Guide for the 80's.* Homewood, Illinois: Dow Jones-Irwin, 1980. More than a job-hunt book. Presents practical tips for developing career strategy, analyzing job change needs, and evaluating prospective employers.

Schein, Edgar H. *Career Dynamics: Matching Individual and Organizational Needs.* Reading, Massachusetts: Addison-Wesley Publishing Co., Inc., 1978. This book studies the complexities of career development from both an individual and an organizational perspective. Good discussion of changing needs throughout the adult life cycle and the interaction of work and family.

CHAPTER 7

Identifying Your Marketable Skills

The greatest pleasure in life is doing what people say you cannot do.

Walter Bagehot

OVERVIEW

You are hired for the contributions you make to an employer. But you are capable of different kinds of contributions, all based on your skills. The skills to "sell" are those you use best and enjoy most. By identifying your most important skills you can select jobs that ensure personal satisfaction and success. This chapter explores:

- The several categories of skills you possess
- How skill awareness increases job mobility options
- Some methods for identifying your own most important skills
- How to use skill awareness in résumé writing and job hunting

AVOID JOB MISMATCHES

Ever been in a job you didn't enjoy or couldn't handle? These mismatches occur when the skills needed for the job aren't your strong suit, and when your best skills go unused. Consider those jobs you liked and performed well. Chances are, they called for skills you were most comfortable with. Awareness of the skills you enjoy using and use well can eliminate job mismatches and increase your job mobility.

I ask my career planning workshop participants to make a list of all their work skills, knowledge, and capabilities. The typical list con-

79

tains a couple of dozen items. But as participants further examine their education and work experience, their lists expand to a few hundred skills.

The skills resource base you accumulate with experience is the basis for career mobility. Most of these skills are transferable to a variety of jobs.

IDENTIFY YOUR MULTIPLE SKILLS

I define the term "skills" to include talents, traits, abilities, strengths, knowledge, and other personal attributes which comprise your unique "capability package." These can be grouped in three skills categories—technical, functional, and self-management. Let's examine each in turn.

Technical Skills

Technical skills make up the job-specific aspects of work and are easy to identify. They include the specialized vocabularies, experimental methods, problem-solving techniques, and analytic procedures which are the warp and woof of work. You acquire technical skills through formal study and job experience.

Technical skills are most prone to obsolescence as new techniques supplant the old. Electronics engineers have seen their technology advance from the vacuum tube to the transistor to the microprocessor. To remain at the technical front line requires maintaining and updating technical skills throughout your career as the technology changes.

Technical skills are vital to job performance, especially in early career years when your experience is limited. Because of their specialized nature, not all these skills transfer to different work settings. The technologist well-grounded in the fundamentals can readily acquire the job-specific skill requirements if the new job is similar.

When people are asked to define their skills, the usual response is a list of technical skills used in the current job. I recently reviewed the résumé of a 40-year-old computer systems analyst. It described all the programs he had developed, and was written in esoteric technical language understandable only to another computer expert.

I asked him to describe his role in these projects. It became apparent that he had major responsibility for interviewing top management to identify information needs, defining system requirements, analyzing alternative configurations, supervising the project team, developing the system, training client staff to operate the system, and so forth.

Yet none of these important skills appeared on his résumé. He was interested in becoming director of systems management for a large firm, but the résumé described only his programming experience. As written, the résumé gave him no competitive advantage over a 25-year-old systems analyst. He ignored the essential skills his 20 years of experience provided which would qualify him for the position he wanted.

Technical skills are only a portion of what most jobs require. Their importance decreases at higher organization positions and greater responsibility levels. Don't sell yourself short by limiting your capabilities to technical skills.

Functional Skills

Dr. Sidney Fine, author of the "Dictionary of Occupational Titles," is an expert in functional job analysis. His voluminous dictionary analyzes the requirements of virtually every job category. Dr. Fine analyzed several thousand individual jobs to identify their generic requirements. He then distilled these into three major categories—data, people, and things—with several functions in each.

He arranged these into a skills hierarchy, starting with the less complex skills (at the bottom) through more complex skills (at the top) and ended up with the diagram shown in Figure 7–1. Each successive functional skill typically includes all the skills ranged below it. If separated by a comma, the skill functions are on the same level and separately defined. If a function is hyphenated, it is a single skill. Skills under Precision working, Manipulating, and Handling are special cases involving machines and equipment, hence indented.

If you can perform the "higher skills" in each category, you can also perform the lower skills from which the higher skills are drawn. For example, the art of synthesizing data also involves coordinating, analyzing, computing compiling, copying, and comparing data. Mentoring includes the ability to negotiate, supervise, instruct, and so forth.

Functional skills are transferable. Unlike job-specific technical skills, functional skills easily transfer across different job categories. The person able to synthesize data in one technical field can do the same in much different fields.

Jobs using lower-level skills are easily described (and generally narrow); those calling for higher-level skills are not as easily described and rely much more on employee discretion.

As an example, the research manager directing a small experimental research team is doing more than applying this technical knowledge to crack a tough technical problem. He is developing and using a wide range of transferable skills—organizing, motivating, and managing a building project team; planning, monitoring, and controlling milestones and schedules; formulating, testing, and verifying experimental hypotheses; collecting, organizing, and analyzing data; and so forth.

Figure 7–1 Three categories of functional skills

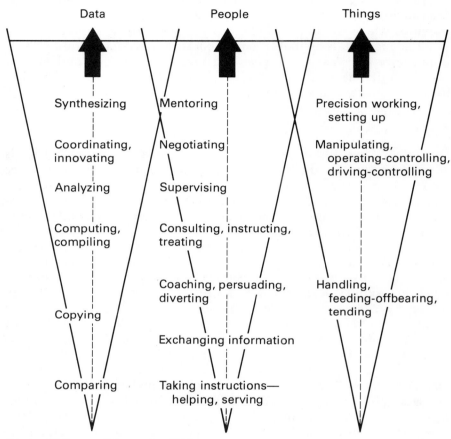

Source: Dictionary of Occupational Titles

These same skills are required in thousands of other jobs. And they extend far beyond research work to jobs in many other industries, organizations, and departments. Developing functional skills is the key to increasing job mobility. An example: Albert Feiner had ten years experience designing commercial satellite ground stations. In the mid-70's, the marriage of two technologies—satellites and education—led to satellites for providing educational services to remote and dispersed populations in Alaska and Appalachia, and in developing countries. Today, Feiner is a leading authority in applying satellite technology to education programs in developing countries. He transplanted his accumulated skills and knowledge of a specific technology to a new arena. He had no experience in education prior to taking on this job, but had solid transferable experience in planning and managing complex programs to complement his technical knowledge.

Self-Management Skills

This third skills category enables people to respond appropriately to various circumstances. Self-management skills are formed in early development and include such personality characteristics as confidence, enthusiasm, poise, persistence, sense of humor and similar attributes.

Self-management skills are on constant display. "Technically he is a whiz, but it's impossible to work with him." "It's a pleasure having her on the team, her enthusiasm for the project is contagious." "His only problem is that he just doesn't listen." Such skills are often the "make or break" factor in careers.

Self-management skills let you choose from your full "inventory" of possible behaviors those appropriate to the situation at hand. Discretion is required in using these skills. Inability to use positive skills moderately can be harmful, as when confidence leads to arrogance, initiative leads to overbearance, and assertiveness leads to manipulation.

DEFINE YOUR MULTIPLE SKILLS

Every person reading this book is capable of identifying between 250–500 different skills, provided you invest sufficient effort "mining" your experiences. To identify these, make a list of your past "peak experiences," work and nonwork experiences which gave you particular satisfaction. Then ask yourself what skills and traits were used in each case. (These skill identification methods were pioneered by Bernard B. Haldane, *Career Satisfaction and Success*, *A Guide to Job Freedom*, and by John C. Crystal and Richard N. Bolles, *Where Do I Go From Here With My Life?*. Haldane first coined the term "motivating skills" used in this chapter.) Each peak experience may contain 10, 15, 20 or more of your most important skills. An example:

> "It was a mess and everyone was confused about what to do. We had to deliver the proposed high-yield seed technology research plan in five days. But it was impossible to get started because of different opinions on technical approach among team members. The problem was really one of personality clash. On a one-to-one basis, I met with the key people to get their ideas. I also listened and let them talk out the personality issues. Then during a team planning meeting, I recommended a compromise approach which included ideas from all the key people. Because time was tight, I had prepared a PERT network schedule for putting together the report, identifying when various written products were needed, review points, report production, and so forth. We hashed it out and were able to get started. We all pitched in and the research plan was finished in time. I feel good about taking the initiative and resolving what was both a personality and technical problem."

Many skills are apparent in this example, including:

- In-depth understanding of high-yield seed technology
- Knowledge of PERT network scheduling techniques
- Ability to inspire trust in others
- Ability to resolve differences and reach compromises
- Skill at developing technical research plans
- Ability to organize multi-disciplinary project teams
- Ability to work well under pressure
- Willingness to take initiative
- Personal tact and diplomacy

Note that these skills fit all three categories. Only one—the understanding of high-yield seed technology—is job-specific. The remainder are transferable functional and self-management skills.

SUMMARIZE YOUR MOTIVATING SKILLS

If you pick a dozen of your own peak experiences and define a few hundred skills, the question then becomes how to *use* such an unwieldy number of miscellaneous items. The answer is to compress them into a power-charged summary of your *motivating skills*—what you do best and enjoy most. This is a sort and sift process of combining similar skills into groups and gradually condensing, until you have a summary that excites you and catches the eye of potential employers because you demonstrate the perfect skills to solve their problems. An example of such a summary:

> Proven proficiency in solving a diverse range of technological problems. Adept at design analysis and research. Can readily assimilate new areas of technology. Effective at quickly grasping the overall problem and identifying solution methods. Approaches problem solving both theoretically and intuitively and can deliver both academic or practical solutions. Accomplished at written and oral presentations requiring organization of highly technical material. Works well with management and specialists at all organization levels.

One of the toughest career planning tasks is writing a job objective which "fits" what you want. It is easy to make a yes/no decision if a potential job is described in detail, but harder to specify an objective ahead of time. As a result, most job objectives are either so broad they are meaningless or so narrow they are restrictive.

Skills identification overcomes this difficulty. Defining your motivating skills is the key step. To these, you need only add a *job function* and *organization type* to create an objective which multiplies your possibilities for finding the right situation. Figure 7–2 summarizes the process.

The point should be clear: you have developed a rich and varied skills base in your work and life experience. These skills are the raw ingredients of career mobility. Awareness of skill transferability is the best way to overcome the limitations of discipline, educational, and organization "tags."

My own career experience is a good example of career mobility through building and using transferable skills. My key skills are planning and organizing, a highly transferable technology that has led to diverse jobs—managing an evaluation of the U.S. air traffic control system, planning agriculture programs in developing countries, analyzing the research planning methods of the Environmental Protection Agency, and organizing career planning systems in large companies. Despite the diversity of these assignments, they were all based on solid planning and organization skills.

For a good example of skill transferability, see the résumé in Figure 7–3. This man is obviously qualified for many different jobs. (Please

Figure 7–2 Developing a skills-oriented job objective

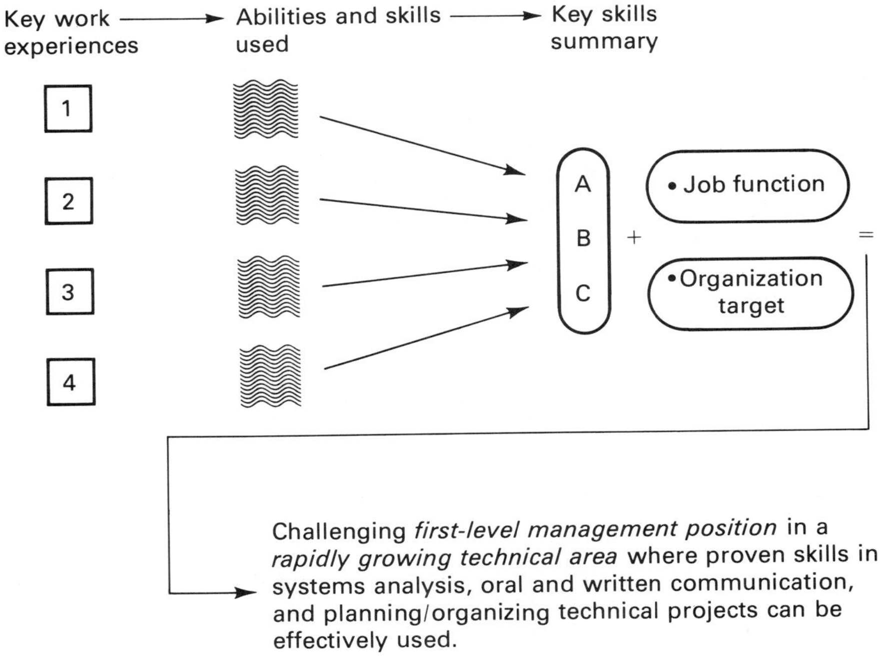

Challenging *first-level management position* in a *rapidly growing technical area* where proven skills in systems analysis, oral and written communication, and planning/organizing technical projects can be effectively used.

Figure 7–3   Résumé of Robert J. Martin, Jr.

Key experiences and skills

- *Chief Executive Officer*   Supervised small, non-profit organization.
- *Budget and Finance Director*   Responsible for overseeing expenditures.
- *Purchasing Manager, Head Buyer*   In charge of acquisitions, ordering of goods and services, inventory control.
- *Negotiator/Arbiter*   In charge of labor disputes and conflict resolution between management and subordinates.
- *Ombudsman*   Oversaw impartial reporting of complaints to other management personnel.
- *Maintenance Foreman*   Administered and executed necessary maintenance and repair procedures to physical plant.
- *Plant Designer*   Established design criteria and executed necessary maintenance and repair procedures to physical plant.
- *Travel Planner/Coordinator*   Responsible for all aspects of individual and group travel plans both inside and outside of U.S.
- *Transportation Director*   Arranged for, acquired and helped maintain and operate various kinds of transportation equipment for both short- and long-haul use.
- *Activity Coordinator*   Responsible for planning and executing activities for individuals, small and large groups.
- *Educational Coordinator*   Administered and oversaw planning of educational programs and classes.
- *Teacher/Instructor*   Designed and taught numerous subjects, lessons, courses and activities.
- *Medical Administrator*   Procured and oversaw health program and administered benefits of same to management and subordinates.

He acquired these skills during a four-year stint as a house-husband!

Adapted from an article by Robert J. Martin, Jr., in *The Washington Post*, June 15, 1981. Reprinted by permission.

resist your temptation to skip to the bottom to discover how he acquired his capabilities.)

The demand for skills also follows the life-cycle curve described in Chapter 2. Not all your skills will be equally useful in your career future; some will have little value, others will be in high demand.

Focus your new-skill development on technologies that have a high future demand probability. This requires looking at how your field is changing and the projected skill needs of the organization. Begin *now* to develop skills to enhance future opportunities.

SUMMARY OF KEY POINTS

- The basis of career progress is demonstrating that you have the skills to help the organization succeed.
- Don't define your skills only as technical skills. Functional skills (people, data, things) and self-management are especially important and highly transferable.
- Transferable skills provide job mobility. It takes some digging to identify these, but it pays off in self-understanding and increased career flexibility.
- Of all your skills, those which motivate you are most important. Your motivating skills are the ones you used in past situations which gave you special satisfaction.
- Skills demand follows the life-cycle curve. Develop new high demand competencies to replace those with static or declining demand.

QUESTIONS TO CONSIDER

1   What are your primary skills? See how long a list you can make of technical, functional, and self-management skills.

2   Which of these do you want to use in future jobs? Which do you not prefer to use?

3   What portion of your knowledge/skills base do you consider transferable? What portion is job-specific?

4   What can you do to gain more transferable skills through the work you have now (e.g., get on project task forces, take on new assignments)?

5   As you project your job/career future within your current organization, which of your skills have a high probability of being needed and which have a high probability of *not* being needed? Do you have an appropriate balance for the future?

RECOMMENDED READINGS

Bolles, Richard N. *The Three Boxes of Life*. Berkeley, California: Ten Speed Press, 1978. A comprehensive treatment of life/work planning, written with a light-hearted twist and whimsical drawings. Bolles' earlier book, *What*

Color Is Your Parachute?, is a classic in the field. Both discuss the topic of skills identification in depth.

Rinella, Richard J. and Robbins, Claire C. *Career Power! A Manual for Personal Career Advancement*. New York: AMACOM, 1980. Good treatment of skills identification, followed by résumé-writing, job-hunting, and interviewing techniques.

CHAPTER 8

Choosing from Major Career Directions

If you don't know where you're going, any road will get you there.

The Koran

OVERVIEW

Your career requires decision making and choosing among alternative directions. This chapter explores the key career decisions faced by engineers and scientists, including:

- Within a technical field, should you generalize or specialize? How and in what areas?
- Should you pursue a technical or a managerial career focus?
- What sector (business, government, academic) should you work in?
- If management is of interest to you, would you be happy as a manager?
- Is running your own business a realistic option?
- What other alternatives are open to you?

MAJOR CAREER DECISIONS AND DIRECTIONS

The career patterns of individual scientists and engineers vary remarkably. The combination, timing, and sequence of individual sector/industry/company/functional job decisions make each career pattern unique. Some people enjoy one career through retirement, confining job changes to the same industry, function, or technology. Others succeed in two, three, or more highly different careers. Whatever your career pattern, it is defined by the decisions you make. Whether or not you decide consciously, decisions get made.

Some persons make key decisions early and easily. Others require more time and work experience. Still others make tentative decisions and later change their minds. Not all decisions need be made at once, nor is any decision irreversible. The best strategy may involve several decision-making cycles, each built upon accumulated experience and greater self-insight.

Decision-making theory states that you have made a decision only when there is a cost in changing your mind. A good strategy lets you pursue several alternatives simultaneously and keep your options open. If you are a bench scientist aspiring to technical management, you can sharpen your management skills while pursuing your research specialty. As long as you do not permit your technical skills to erode in anticipation of becoming a manager, you create additional future options, both technical and managerial, without pre-empting any potential choices.

WHO HIRES: EMPLOYMENT BY SECTOR AND ACTIVITY

According to the National Science Foundation, some 78% of the nation's engineers and 45% of the nation's scientists work for business and industry. Educational institutions are the second largest category, hiring about 28% of the scientists and about 4% of the engineers. But educational jobs have steadily decreased as enrollment declines since the late 60's have forced faculty cut-backs.

Nearly one out of every ten scientists and engineers works directly for Uncle Sam, the third largest employer. Engineers are 44% of the total, the largest group of technical specialists employed by the federal government. Other major employers are state and local government, nonprofit organizations, and the military. Figure 8–1 displays employer types; Appendix I includes more detailed breakdowns by employer and specialty type.

Mobility barriers between different sectors are low; many identical jobs are found in all sectors. But the work climate varies by sector as well as by organization. The important part of sector "fit" is selecting an environment compatible with your work style.

You are likely to perform many different types of work during your career. Figure 8–2 illustrates the primary work activity of employed engineers and scientists; Appendix I includes a more detailed breakdown of employment activity by field and sex.

Figure 8–1 Employed scientists and engineers by
type of employer

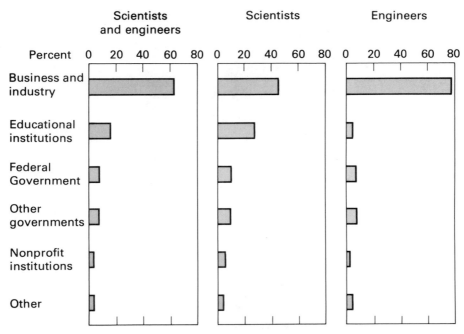

Source: National Science Foundation

R & D is the largest single category of scientific and engineering employment. It accounts for 28% of technical employment; an additional 9% work in R & D management and administration. R & D employment is highly sensitive to federal funding for energy, space, and defense programs.

Management is the primary work activity of 25% of scientists and engineers. Scientists who manage are likely to manage R & D; engineers manage in more general fields. Engineers are more likely to become managers than scientists. Engineering managers tend to hold bachelor's degrees, rather than master's or doctorates.

Some 9% of scientists and engineers teach. Scientists are more inclined to teach than engineers: 17% of the scientists versus 2% of the engineers teach as their primary professional activity.

The barrier to switching work functions is skill-related. You can reduce this barrier by developing *transferable skills,* as discussed in Chapter 7.

The rest of this chapter explores the major functional options and provides guidelines for making your choices.

Figure 8–2 Employed scientists and engineers by
primary work activity

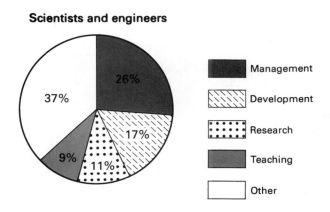

Scientists and engineers

- Management
- Development
- Research
- Teaching
- Other

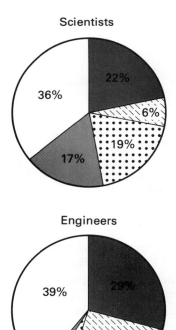

Scientists

Engineers

Source: National Science Foundation

PURSUING TECHNICAL CAREERS

Technical career paths provide the intrinsic satisfaction of solving challenging problems, and many scientists and engineers successfully remain on the technical front line throughout their career.

But the same factors that make technical careers exciting also create risk. Technologists face a special challenge: how to cope with the explosion of new technical knowledge and the splintering of fields into esoteric sub-disciplines. The key issue is how to straddle the spectrum of specializing versus generalizing.

Specializing or Generalizing

To explore your options, imagine that all the world's knowledge could be displayed on a grid. Consider the horizontal dimension to be *breadth* of knowledge; the vertical dimension to be *depth*. Every person has some combination of knowledge depth and breadth.

Now consider two opposite strategies for coping with the information explosion: You could respond to knowledge growth by focusing on narrower fields, mastering a smaller and smaller subject area, until you know everything about nothing! The problem with being an "overspecialized specialist," of course, is that you would have total depth but no breadth. This knowledge profile is highly vulnerable to technology changes or layoffs.

The opposite strategy would be to develop some understanding of virtually everything but acquire no real depth. You would learn less and less about more and more, until you finally knew nothing about everything! As an "overgeneralized generalist" you would fit few environments because you would have no skills differentiation or strength.

With these two extremes in mind, here are two practical profiles which may suggest a strategy for you. *Focused specialists* build expertise around a few closely related technologies. They avoid the problem of overspecialization by remaining conversant in several related fields. They minimize career risks by staying abreast of related technology where the "incremental cost" (time and energy) of remaining current is not too high.

For example, electronics engineers expert at computer disc storage techniques who also stay abreast of other storage techniques can analyze their overlap with discs, predict their usefulness, and so forth. This modest broadening of a specialty increases their ability to move into related storage technologies should they so choose.

Concentrated generalists have a broader knowledge base. They work to create new combinations from this knowledge and use their transferable skills to build bridges to new fields.

My own career history demonstrates this last profile. Through education, work experience, and self-instruction, I have acquired an understanding of technical research, program management, and human resource development. This book draws ideas from these three different areas and "transplants" the concepts to the field of career planning.

How Specialists Can Adapt

Technologists who have narrowly specialized for many years face a different challenge. They may work with younger colleagues with more current education who can do the same technical job. If they enjoy the work and perform well, they needn't change. But they can also capitalize on their years of experience by making a different type of organization contribution.

One approach is to *generalize* within their area of competence. This involves new roles in which accumulated knowledge and experience can pay off. Possible roles include increased dealings with the external environment, responsibility for developing other people, generating new technical ideas, or serving as a task leader or project manager.

A second approach is moving toward *applications* functions which require less up-to-date technical expertise. Development engineers, for example, can move to manufacturing, marketing, technical support, and so forth.

Changing roles and taking on greater responsibility needn't require managerial or administrative work.

Dual Career Ladder

Most technical organizations maintain a "dual career ladder," with one career branch advancing in a technical direction, the other in a management direction. The dual ladder is a way to recognize, reward, and promote technically skilled people whose interests do not lie in management. Figure 8–3 illustrates the dual ladder maintained by the Westinghouse Company.

Reprinted by permission of Johnny Hart and Field Enterprises, Inc.

Figure 8–3　The dual career ladder

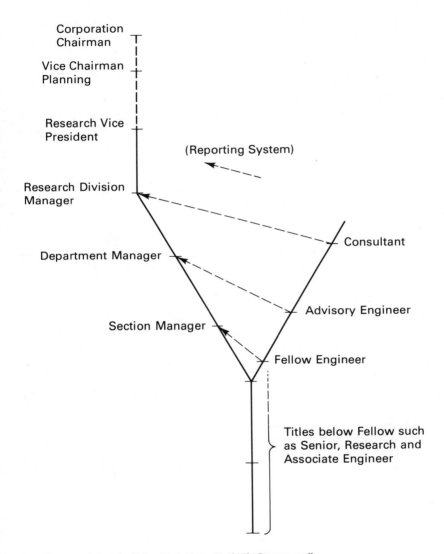

From "Maintaining Staff Productivity as Half-Life Decreases,"
by C.W. Kimblin and W.E. Souder, *Research Management*,
November, 1975. Reprinted by permission.

　　　Technical ladders combine the benefits of vertical growth with
the opportunity to stay technically involved. But there are drawbacks. In some organizations, ineffective general managers are moved
laterally to the upper ends of the technical ladder. The technical

management ladder generally stops short of the highest organization positions.

Perhaps the toughest decision for scientists and engineers is whether to remain with the technical action through their career or move into management. This question must be faced once, or many times. But however one decides the management versus technology question, circumstances may act otherwise. Not every engineer who aspires to management will be perceived as capable or have the opportunity. Likewise, many scientists and engineers who are content to remain in technology find themselves "drafted" into management positions. To manage or not, that is the question, and the next section may help you answer.

TAKING THE MANAGEMENT ROUTE

Far too many competent, satisfied scientists and engineers become incompetent, dissatisfied managers. Moving into management usually requires giving up a primary source of career satisfaction—involvement with the technical action.

Managers address "system-centered" issues, as opposed to the "problem-centered" challenges of technologists. Management demands different perspectives which some people lack or are uncomfortable using. To successfully move from technology to management requires scientists and engineers to alter their modes of thinking and problem solving.

Engineering and the sciences are *convergent* disciplines. For any problem, there is usually one or more rational solution, achieved through analysis, calculation, experimentation, and testing. The best solution can be *proven* by equations and principles.

Management, however, is a *divergent* discipline. The best approach cannot be proven *a priori;* there may be no solution. The problem-solving

domain is broad, the variables difficult to control. Managers deal with the uncertainty of people rather than the certainty of equations, data, and equipment.

Management, however, can yield psychic and financial rewards matched by few technical jobs and is a vehicle to the very top—some 20% of company presidents have progressed from technical jobs to technical management to top management.

Difficulties in Becoming A Manager

The U.S. government studied engineers and scientists promoted to management at the National Institutes of Health (NIH) and the National Aeronautics and Space Administration (NASA). It examined 16 skills to determine which were likely to be sources of problems for the scientist or engineer who moves to management. The 16 skills were:

1 *Fundamental technology*   Is well-founded in the fundamentals of a specific field

2 *Application of techniques*   Knows how to apply techniques

3 *Knowledge in related areas*   Has professional knowledge in areas related to specialty

4 *Operating within organizational system*   Understands organizational goals, structure, relationships, and procedures

5 *Operating within financial system*   Understands relevant budgeting, cost estimating, and cost-control techniques or procedures

6 *Operating within personnel system*   Knows the informal means (and restrictions) applicable to the full range of personnel activities from recruitment through separation

7 *Recognizing, coping with environmental factors*   Is familiar with and can deal with constituent or professional-group interests, interagency problems or relations, interested officials in other component organizations within the agency, organizational politics

8 *Communication of ideas*   Knows how to communicate ideas

9 *Working with diverse people*   Can work with people of diverse ability, type, and temperament

10 *Coordinating group effort*   Has ability to coordinate and facilitate group efforts, negotiate

11 *Leadership style*   Possesses a leadership style that draws positive responses from subordinates

12 *Generation of confidence of superior*   Can gain confidence of superior

13 *Integrative ability*   Has ability to perceive and assess relationships

14   *Problem solving*   Can identify and define critical issues to develop potential solutions

15   *Decision making*   Possesses decision-making ability

16   *Creative thinking*   Is capable of creative thinking

The study concluded that the problems faced by the scientist/engineer turned manager were not with the planning and organizing aspects of management—most technical jobs include a healthy dose of these. Rather, the difficulty rested with the interpersonal and political interaction required with other elements of the organization. Few engineers or scientists are trained to do this well. Figures 8–4 through 8–6 summarize these findings.

Prepare Yourself to Manage

If you aspire to management, prepare yourself to succeed before you take on the position. Your organization may have management training programs; most local colleges have programs. If these resources are unavailable, you may have to be more inventive, as this person was:

> "I created my own management development program. My mentor helped me plan a strategy of self-study and expanded activities on and off the job. I asked my supervisor to help me gain management experience, and he was delighted to delegate some of his tasks.
>
> "I stopped watching Monday night football and started reading good management books, beginning with a college text and moving on to Peter Drucker and others. But my best management training came when I volunteered for my church's spring picnic committee. This involved all facets of management—motivating others, budgeting, advertising, negotiating, planning, and scheduling. I was thinking and acting like a manager. Not long after, I was offered a management job, and believe me, I was prepared."

If management interests you, candidly rate yourself on the 16 key skills. Which are you comfortable with, which need sharpening? What might you do, on the job and off, to improve these skills?

STARTING YOUR OWN BUSINESS

You may sometimes fantasize about the joys of running your own business—setting your own hours, mastering your destiny, enjoying untold prestige and wealth. And a half million brave souls actually set up a corporation each year.

Figure 8-4 Percentages of NASA engineers regarding various management skills as most likely to be sources of difficulty, for a specialist moving into management

Skill	Percentage reporting each skill as a likely source of difficulty			
	Bench (N = 31*)	Supervisors (N = 49*)	Managers (N = 51*)	Senior managers (N = 30*)
Fundamental technology	—	—	1	—
Application of techniques	—	—	4	3
Knowledge in related areas	10	4	8	7
Operating within organizational system	26	29	37	37
Operating within financial system	32	24	41	37
Operating within personnel system	36	24	33	23
Recognizing, coping with environmental factors	32	26	35	50
Communication of ideas	10	12	26	17
Working with diverse people	42	39	39	50
Coordinating group effort	19	16	22	23
Leadership style	29	24	24	23
Generation of confidence of superior	—	—	6	3
Integrative ability	—	8	8	3
Problem solving	—	4	2	—
Decision making	13	16	14	17
Creative thinking	—	2	2	3
None a source of difficulty	—	8	—	3
Depends on person replaced or position taken	—	—	2	—

*N = number of individuals in sample

Note: Percentages add to more than 100 because of multiple answers.

Source: James A. Bayton and Richard L. Chapman, "Transformation of Scientists and Engineers into Managers," NASA Report SP-291

Figure 8–5 Percentages of NIH scientists regarding various management skills as most likely to be sources of difficulty, for a specialist moving into management

Skill	Percentage reporting each skill as a likely source of difficulty			
	Bench (N = 30*)	Supervisors (N = 51*)	Managers (N = 50*)	Senior managers (N = 30*)
Fundamental technology	—	2	2	3
Application of techniques	3	—	4	3
Knowledge in related areas	3	2	2	—
Operating within organizational system	55	46	53	73
Operating within financial system	38	40	35	73
Operating within personnel system	45	48	31	60
Recognizing, coping with environmental factors	38	26	31	53
Communication of ideas	3	2	6	7
Working with diverse people	21	16	27	30
Coordinating group effort	31	14	29	17
Leadership style	28	22	10	10
Generation of confidence of superior	7	4	6	—
Integrative ability	—	6	4	—
Problem solving	3	12	8	—
Decision making	3	20	6	17
Creative thinking	3	4	4	3
None a source of difficulty	7	6	8	—
Depends on person replaced or position taken	7	—	2	—

*N = number of individuals in sample

Note: Percentages add to more than 100 because of multiple answers.

Source: James A. Bayton and Richard L. Chapman, "Transformation of Scientists and Engineers into Managers," NASA Report SP-291

Figure 8–6 Percentages of NASA scientists regarding various management skills as most likely to be sources of difficulty, for a specialist moving into management

Skill	Percentage reporting each skill as a likely source of difficulty			
	Bench (N = 35*)	Supervisors (N = 49*)	Managers (N = 50*)	Senior managers (N = 33*)
Fundamental technology	3	2	2	6
Application of techniques	—	2	—	—
Knowledge in related areas	6	8	6	3
Operating within organizational system	34	31	36	52
Operating within financial system	43	33	26	58
Operating within personnel system	29	31	30	48
Recognizing, coping with environmental factors	29	31	34	48
Communication of ideas	6	16	16	9
Working with diverse people	17	29	34	46
Coordinating group effort	26	24	26	18
Leadership style	34	31	16	18
Generation of confidence of superior	6	2	6	6
Integrative ability	6	2	2	9
Problem solving	3	2	—	—
Decision making	23	12	18	6
Creative thinking	3	—	—	—
None a source of difficulty	6	2	2	—
Depends on person replaced or position taken	—	—	—	—

*N = number of individuals in sample

Note: Percentages add to more than 100 because of multiple answers.

Source: James A. Bayton and Richard L. Chapman, "Transformation of Scientists and Engineers into Managers," NASA Report SP-291

You may be inspired by reading stories about companies such as Apple Computer, whose founders parlayed a modest investment into a $300 million business in five years. You don't read about the three out of five businesses which fail within five years.

The cost of running your own business can be high, not just the cash investment, but the psychic investment and tradeoffs. The time and energy required to make a new business succeed exact a high toll on spouse, children, and personal concerns.

The glamorous illusions fade during the uncertainty and stress of start-up. The reality means working 80-hour weeks, fending off creditors, emptying your own trash can, sweating to meet payroll, and discovering—again—that there's no cash for your rent.

The reasons most entrepreneurs fail is not because they lack technical skills. Rather, they are deficient in the other critical (but less satisfying) requirements of *marketing* their services, *financing* their enterprise, and *managing* their business.

Statistically, your chances of entrepreneurial success are greater if you fit the following profile:

• Hold a master's degree, rather than a bachelor's or doctorate
• Demonstrated entrepreneurial tendencies while young
• Are the son or daughter of a self-employed parent
• Have a psychologically supportive spouse
• Start your first venture before reaching age 40

There are three ways to minimize your risk. One is to first work for an enterprise similar to the one you wish to enter: climb the learning curve on someone else's payroll. The second is to buy into an existing small business which has weathered the critical start-up period. The third is to begin part time, perhaps working from your home to reduce overhead.

If enterpreneurial blood runs deep in your veins and you make the plunge, do so with a sound strategy. The best of luck. You are bucking the odds, but you may become the next small business success story.

SECOND CAREERS

Talented individuals can succeed in two, three, or more highly different careers as lifespans lengthen, opportunities multiply, and society accepts more diverse life styles. The notion of "one life, one career" is obsolete.

A second career—an occupation totally different from one trained for and worked at for many years—arises for different reasons. It may arise when you conclude your present career is too restrictive. For some, second career decisions correct earlier mistakes and permit "following a dream." For others, the second career is the logical answer to the need for a dramatic life-style change.

Some second careers are deliberate; others are forced. Many talented aerospace engineers hit by the cutbacks of the late 60's, for example, chose second careers by necessity when options ceased in their preferred fields. Project cancellations, mergers, and reorganizations can force you to consider second careers. This engineer's second career was forced:

> "It all happened so suddenly. One day I was top technical banana in a new product division, the next day out on the street when the company decided it couldn't compete in that business. There wasn't much work in that part of New Jersey, and our original reasons for living there had changed. We packed up the car and headed south, with no particular destination in mind.
>
> "We drove along the coast, through Virginia, Carolina, Georgia, Florida. I told my wife to say 'stop' when she spotted a town where she'd like to live. She said 'stop' in West Palm Beach, so we checked into a hotel and I started looking. We lived for two months on credit cards until I found work, as an assistant apartment manager. Not great money, but it included a place to live. Today, I manage three large apartment complexes. I'm doing better than ever, my health is improved, and there is plenty of time for fishing. And we live less than a mile from where my wife said 'stop'! It was rough going for awhile, but if I knew it would turn out like this, I would have changed long ago."

Chosen second or third careers can provide challenge, fulfillment, and satisfaction. But consider carefully. Abandoning the comfortable familiarity of your current profession can be traumatic. Second careers burn professional and personal bridges to the past. Colleagues and friends seldom understand such dramatic self-chosen changes and may consider you a dropout. Before acting, get a "reality check" from a mentor, close friend, or colleague.

Most people who take this path describe no regrets, after the new life style gets on track. When I quit a comfortable, secure job to start a management consulting enterprise in a different field, my change in self-esteem over the next few months resembled the shape of a square-root symbol ($\sqrt{}$). After making the break, I plunged into uncertainty and doubt over whether the decision was right, but after some difficult months, my personal satisfaction rose to a new level.

SUMMARY OF KEY POINTS

- Career mobility between different employing sectors requires *value* flexibility; between different functions, *skill* flexibility.
- Technical career risks can be overcome by awareness of the prob-

lems of too narrow specialization. Your technical career experience can be exploited by finding ways to add to organization value.

- The key difficulty scientists and engineers face as managers is adjusting from "problem-centered" responsibilities to "system-centered" roles. The latter requires interpersonal and political interaction skills, but technical jobs seldom provide this experience.

- Reduce the risk of failing if you run your own business by thoroughly understanding what it requires, developing a sound marketing plan, and keeping overhead low.

- Second careers, whether chosen or forced, provide interesting opportunities to start over career-wise.

QUESTIONS TO CONSIDER

1   What are the major decisions you have faced in your current career?
2   What are the major decisions which remain to be made in your current career?
3   What second (or third) careers seem intriguing to you? If there were no "switching costs," which would you choose? Why?

RECOMMENDED READINGS

Verralin, Charles H. *Management and the Technical Profession*. Houston: Gulf Publishing Co., 1981. A useful compilation of 58 articles by 34 authors on leadership, self-management, communication, problem solving, and related topics. These articles have appeared over the years in *Hydrocarbon Processing*. Equally helpful to nonmanagers.

Bayton, James A. and Chapman, Richard L. *Transformation of Scientists and Engineers into Managers*. Washington, D.C.: NASA's Scientific and Technical Information Office, 1972. Extensive research report on current and prospective technical managers and the problems they encounter in making the transition.

CHAPTER 9

Making Effective Career and Life Decisions

Not to decide is to decide.
Anonymous

OVERVIEW

Implementing career strategy involves countless decisions. Most decisions are minor, but a few major decisions deserve your best thinking. This chapter provides practical tools for making important decisions and discusses:

- A logical process for making better decisions
- Clarifying the decision criteria you use
- How to create more alternatives to choose from
- The psychology of decision making

CRITICAL DECISIONS DESERVE RIGOR

Your life history centers around major decisions you make and events that occur. You are who and what you are today because of decisions made in the past. You decided what to do after high school, what and where to study, whether and who to marry, which jobs to take and when to change, and much more.

Who and what you will be in the future depends on decisions yet to be made. Your life is also shaped by events, planned or unplanned, fortunate or unfortunate. Discouraging life events—a serious illness, loss of a loved one, job layoff—shape our lives as do the unexpected opportunities and serendipitous blessings.

While not all life events can be controlled, you can control your decisions. You'll make countless decisions over your life, the majority small and inconsequential. But a few critical decisions deserve careful consideration to make the best possible decision. An example of how one individual analyzed a key job decision:

> "It was the best job offer I ever received. The unexpected offer came through an old friend and it was perfect for me—great company, good money, big responsibility. Just one drawback—it was at the other end of the country. As I mulled it over, it became apparent how much we'd have to sacrifice for me to take it. My family has built strong friendships in the community. My kids would have to give up their swim team and change schools. My wife's career is getting on track. I've just been elected to the local neighborhood council. I'd have to give all that up, plus my Kiwanis Club, Wednesday night bowling, and other things which I enjoy. Sure, we could find all that in the new location, but it would take years to get back in the groove. I concluded that the sacrifices were just too great, so I turned it down. Funny, but five years ago, I would have jumped at this opportunity."

Whether your decision situation arises from out of the blue or through deliberate actions on your part, effective decision making requires both rational left-brain thinking processes balanced by right-brain intuitive thinking. First, let's examine a rational decision-making process useful in many situations.

LOLLY
by Pete Hansen

Reprinted by permission of Tribune Company Syndicate, Inc.

MAKING EFFECTIVE DECISIONS

Decision making is a structured process of choosing the best of two or more alternatives. But it's not always linear. Complex issues involve multiple, interrelated decisions concerning various aspects of the problem.

I find it helpful to view decision making as a process of:

- Clarifying the issue
- Generating possible ideas
- Defining decision criteria
- Evaluating alternative solutions
- Making the final choice

The following sections illustrate these steps and show some practical tools for making better decisions.

Clarify the Issue

Have you ever had a nagging sense that something needs attention, but you weren't sure what? "Issue clarification" sorts out the situation and breaks the problem into manageable parts. Too often people respond to symptoms rather than problems and address the wrong issue. The result is predictable—despite a surface change, the underlying problem lingers. But an issue well defined is half solved.

Consider the example of John Folger, a 31-year-old mechanical engineer in the development department of a large, reputable electrical machinery manufacturing company. Folger described himself as not unhappy, but not happy either. To clarify his concern, I asked Folger to state the issue he would like to improve, beginning with the words "how to." He responded with "how to make my situation more enriching and challenging."

I then asked Folger to restate the issue at least 10 different ways, using the words "how to." His list included:

How to get more pleasure from my work

How to make my work more challenging

How to work with a greater variety of people

How to figure out what I'm good at

How to get more variety in my work

How to find a job where I have more responsibility

How to increase my income

How to get out of a rut

How to eliminate job boredom

How to use more of my unused skills

We talked further. He was proud of his firm, and didn't want to switch companies. But he needed a job change to boost his personal vitality.

As we reviewed his list, I challenged him to synthesize his various definitions into a single statement. He came up with "How to find a more responsible, challenging job with variety and higher income potential without changing firms."

This definition "clicked" for Folger and illustrates the power of the "how to" tool. Defining the issue in many different ways, then summarizing, sharpens your definition.

Generate Possible Solutions

When the issue is well defined, specify possible solutions. The rules of brainstorming apply here—produce a large number of ideas without evaluating, judging, or criticizing.

Folger's list of possibilities consisted of other organizational departments to consider. These were:

Research

Field technical support

Marketing

Administration

Finance

Folger excluded jobs in other companies because he wanted to stay in his company. Had he not decided to stay, his possibilities list would have broadened to include other industries and companies.

Define Decision Criteria

To select from alternatives requires choice criteria to ensure that your decisions satisfy you. I asked Folger to define the criteria his new department should meet, specifying both criteria which are essential and others which are desirable but not essential. He came up with:

Essential Criteria

- Uses my knowledge and expertise in engine generators
- Occasional travel
- Compatible colleagues (with technical backgrounds)
- Income at least equal to present salary

Desirable Criteria

- Chance to acquire new skills
- Out-of-town travel not more than 15% of time
- Chance to work with people outside the firm
- Income potential greater than present salary

His essential criteria are "must-have" and non-negotiable; the desirable criteria are "nice to have" but negotiable.

Note that some criteria have both an essential and a desirable component. Folger sees some travel as essential, but he doesn't want travel to exceed 15%. He wants the job's income potential to exceed his present salary, but it must at least equal his present salary. Essential criteria are firm; desirable criteria can be traded off.

Evaluate Alternatives

A decision-making framework like that of Figure 9–1 helps organize the data needed to evaluate alternatives. This framework supports many types of comparisons, such as different work functions, companies, industries, fields, and specific jobs. Such frameworks don't make decisions, of course, but they ensure that you thoroughly consider all important factors to make the best decision.

Collect evidence to determine how well each alternative meets the criteria. The alternatives must meet the essential criteria to be good candidates; those which don't are eliminated. The remaining alternatives are evaluated further against the desirable criteria. Criteria can be relaxed or tightened to increase or decrease the number of alternatives if the number of options is too small or too overwhelming.

Folger listed his alternatives (including his present department, development). To get information about the other departments, he reviewed written department descriptions and contacted friends in other departments.

The completed matrix showed that Administration and Finance met none of the essential criteria. Research and Development met most but not all. Field Technical Support and Marketing met all his criteria.

Folger continued this analysis by ranking the departments and examining specific jobs in Marketing and Field Technical Support. He also investigated potential openings and application procedures. After he identified several specific jobs of interest, he evaluated them using a similar decision-making matrix.

The decision-making process is iterative and flexible. In comparing alternatives, you may identify additional criteria or modify existing ones. Additional alternatives not originally defined may become apparent.

Figure 9–1   Career decision-making matrix

Decision-making objective: to choose the best: *department in my company*		Alternatives to consider			
		Alternative A: *Development*		Alternative B: *Research*	
Essential criteria ("must have")		Evidence that "A" meets criteria	OK? (yes/no)	Evidence that "B" meets criteria	OK? (yes/no)
Uses my knowledge and expertise concerning engine-generators		*Direct and frequent use*	*yes*	*Direct and frequent use*	*yes*
Occasional travel opportunities		*Not much travel*	*no*	*Travel less than 5%*	*no*
Colleagues have technical backgrounds		*Majority are engineers*	*yes*	*Majority are engineers*	*yes*
Income at least equal to present salary		*Salary structures are similar*	*yes*	*Salary structures are about equal*	*yes*
Does alternative meet all essential criteria?			*no*		*no*
Desirable criteria ("like to have")	Importance	Evidence that "A" meets criteria	OK? (yes/no)	Evidence that "B" meets criteria	OK? (yes/no)
Chance to acquire new skills and knowledge	*High*	*Not too much*	*no*	*Not much— work very similar to mine*	*no*
Travel does not exceed 15% of time	*Low*	*Travel is only 2-5% of time*	*yes*	*Travel less than 5% of time*	*yes*
Contact with people outside company	*Medium*	*Contact is infrequent*	*no*	*Little opportunity for outside contact*	*no*
Income potential greater than present salary	*Medium*	*If get on the "right" project*	*yes*	*Not sure — need more data*	*?*
Analysis and tentative choice		*Meets most of my criteria. Only drawback is lack of travel. This department isn't too bad compared with Administration and Finance. Tolerable but not exciting.* *Third choice*		*Too similar to what I'm doing now. I need a bigger change.* *Fourth choice*	

Alternatives to consider							
Alternative C: *Field Technical Support*		Alternative D: *Marketing*		Alternative E: *Administration*		Alternative F: *Finance*	
Evidence that "C" meets criteria	OK? (yes/no)	Evidence that "D" meets criteria	OK? (yes/no)	Evidence that "E" meets criteria	OK? (yes/no)	Evidence that "F" meets criteria	OK? (yes/no)
My expertise would be a real plus	yes	Knowing the product technically is helpful here	yes	Little chance to apply this knowledge	no	No chance to apply this knowledge	no
Extensive travel to customer site	yes	Frequent travel	yes	Almost no travel	no	Very little travel	no
About half are engineers	yes	About one-third are engineers	yes	Most have business backgrounds	no	Few technical; most from outside	no
Salary structures are similar	yes	Commission system may not match present salary in first year	yes?	Salaries tend to be lower	no	Salaries somewhat higher	yes
	yes		yes		no		no
Evidence that "C" meets criteria	OK? (yes/no)	Evidence that "D" meets criteria	OK? (yes/no)	Evidence that "E" meets criteria	OK? (yes/no)	Evidence that "F" meets criteria	OK? (yes/no)
Would learn about customers business	yes	Would learn proposal-writing, costing, etc.	yes	Not sure what I'd learn	?	Would learn budgeting, cash flow analysis, etc.	yes
Travel averages 25-30%	no	Travel averages 20-25%	no	Very little travel	yes	Little travel	yes
Lots of outside contact	yes	Lots of outside contact	yes	Contact with vendors, supplies	yes	Infrequent contact	no
Potentially—this area is growing	yes	After first year, good potential if I produce	yes	Salaries peak at low level	no	Only if I get an MBA	yes
Good opportunity to learn about many different industries. Will make many contacts. In company growth area — looks good! *First choice*		Travel and outside contact similar to field technical support. May suffer initial income drop, but could do well financially in later years. *Second choice*		Meets none of my essential criteria. Doesn't let me use my strengths.		Meets only one of my essential criteria. "Personality fit" might not work.	

Make the Final Choice

These steps narrow several alternatives down to a final option. In other situations, you may only have one alternative to consider. In either case, the final step is to make a go/no-go decision. The "force-field" analysis tool developed by psychologist Kurt Lewin is ideal for weighing the pros and cons and making a final decision.

In any decision situation there are forces acting for and against the course of action. Force-field analysis helps analyze these forces.

Figure 9–2 illustrates this tool, using a different example. Start by listing forces for and against the course of action. Circle the dominant factors (or use vectors to indicate the strength of each force). If the forces for or against are overwhelming, the decision is easy to make. But if the forces are equivalent, go further to identify steps you can take to increase the strength of driving forces and reduce the strength of restraining forces.

This tool is ideal for testing a possible course of action when you are uncertain. Should I change jobs? Should I stay in my industry? Should I move into management?

Force-field analysis also helps to analyze interpersonal relations and personal growth situations, such as improving relationships with a colleague, resolving a simmering conflict, increasing one's assertiveness, and similar situations.

TRUST YOUR FEELINGS

The most rational decision doesn't always turn out as you planned. Decision making involves uncertainty and risk: the information is never perfect, the outcomes cannot be accurately predicted.

I was once offered a job by an organization with an excellent reputation. The work was appealing; the opportunities were unlimited. On the surface, the job was perfect. But there was something that made me hesitate. I couldn't put my finger on it, but joining that firm just didn't *feel* like the right decision.

Six months later, the firm collapsed. As it turned out, the president was caught using shady business practices. The story made front-page news, and one by one contracts evaporated and the business folded.

My logical analysis said join the company, my intuition said don't. My intuition was right, even though I couldn't clarify why I felt uncomfortable at the time. The reason for listening closely to feelings is that they reflect underlying needs and personal values. These values are hard to define with rational, left-brain thinking. But making decisions compatible with these inner signals is essential for satisfaction.

Figure 9–2   Force-field analysis for making "go/no-go" decisions

Course of action under consideration: Whether to start a part-time consulting business

Forces in favor of action →	← Forces against action
Sense of independence and personal freedom	Potential conflict of interest with present job
Increase income potential	Time conflicts with family and personal interests
Could grow into full-time business, with other employees	Uncertain about my skills in marketing and proposal writing
Would provide stimulation and challenge lacking in my regular job	

I believe the best way to make important decisions is to use rational analytic techniques to identify the best option, then let the option "percolate" before making a final decision. Sleep on it. Give your right brain the chance to work. If your intuition and feelings don't object, the decision is probably correct. But if the decision doesn't feel right, it's the wrong decision. When your brain and gut give you conflicting recommendations, trust your gut feelings.

SUMMARY OF KEY POINTS

- Important life and work decisions deserve rigor and structured analysis. Key decisions should not be taken lightly; invest the time and energy to make good ones. Defining the issue correctly is the starting point. A well-defined problem is half solved.
- Clarify your decision criteria, separating them into necessary and desirable elements.
- Your intuition plays an important role in rational decision making. After the rational analysis suggests the best decision, let your intuition review and endorse the decision. Choices which don't feel right are probably not right.

QUESTIONS TO CONSIDER

1   Reviewing your past, what were the two or three most important decisions you made?

2   What important decisions are you facing now or do you foresee in the near future?

RECOMMENDED READINGS

Byrd, Richard E. *A Guide to Personal Risk Taking.* New York: AMACOM, 1974. Excellent treatment of the psychological dimension of decision making. Shows how to identify creative options and take calculated risks to improve life satisfaction.

CHAPTER 10

Defining and Achieving Your Goals

The journey of a thousand miles begins with a single step.

Chinese proverb

OVERVIEW

Clear goals put you in charge of your life. They motivate and focus your efforts towards desired results. The tips and techniques of goal setting described here apply both to work and life goals. This chapter explains:

- The importance of setting goals and objectives in all life categories
- How to set—and reset—long-, medium-, and short-range objectives
- Analyzing alternatives to select the best path
- How to develop synergystic goal-achievement plans
- Means to monitor progress and measure your success

GOAL SETTING BUILDS VITALITY

Mediocrity is easy to achieve; following the path of least resistance automatically produces this outcome. It takes little effort to drift along without giving serious thought to long-term goals. But by following this route, individuals inevitably discover the ache of aspirations not achieved, the nag of abilities not tested, the frustration of knowing they could have been more had they demanded personal excellence.

People avoid goal setting for different reasons. Many don't know how to start. Others are discouraged by New Year's resolutions repeatedly broken. Still others fear risk and possible failure—without goals,

115

GRIN AND BEAR IT   by Lichty & Wagner

"My lifelong ambition is to someday hitch a ride on the space shuttle."

© 1981 Field Enterprises, Inc. Courtesy of Field Newspaper Syndicate.

you can't miss the mark. Some confine their goals to work, ignoring other important elements. The reasons for ineffective or missing goals go on and on.

Goal setting can provide amazing benefits to your personal vitality and life satisfaction. Even modest goals add special tempo and meaning to your daily routine. An example:

"My goal is deceptively simple: to feel good about myself and others and experience at least a half-dozen small success episodes the first hour of each day. My first success begins when I get up at 7 sharp, my second when I do 25 pushups, my third while I use my shaving time to reaffirm my worth and confidence. My fourth comes during breakfast through upbeat conversation with my wife or kids. While walking to the bus stop I try to acknowledge other people with a smile, a non-verbal gesture, a greeting. I accumulate my half-dozen small victories before I reach the office, and this puts me in a bouyant mood for dealing with whatever the day should bring."

The Goal-Achievement Cycle

Planning, implementation, evaluation—this is the continuing cycle by which goals are set and achieved.

It begins with *planning*—defining your goals and objectives. We all have goals, some vague and distant, others clear and immediate. They fall into several categories of work, family, self; they encompass long-, medium-, short-term, and immediate time frames. A strategic plan combines these into a coherent pattern and sets priorities so that goals can be put into action.

Goals become reality only when implemented. Planning aims at the future; *implementation* occurs now. Goal implementation means using today's 24 hours to accomplish the tasks of the day while also contributing to tomorrow's goals. Implementation requires personal efficiency and good time management; otherwise the demands of the day drive out long-term goal achievement efforts.

Evaluation completes the goal cycle. Periodic progress reviews are needed to refine your goals in light of your experience. The passage of time changes all the variables. You are slightly different than you were yesterday; much different than five years ago. Your values, interests, talents, and goals evolve and merit periodic review.

Changes in the world outside you affect your goals. The assumptions inherent in your strategy change, some radically. The opportunities you identified when setting initial goals change: new possibilities emerge; old ones vanish. Evaluation leads to goal replanning to complete the continuing cycle.

HOW TO SET GOALS

There are many ways to set goals. The simplest way is to take a sheet of paper and brainstorm. List all the things you'd like to do/become/ achieve/experience. Include goals in all areas of importance to you— professional, personal, family, social, cultural, spiritual, and so forth. Sort and refine your list to come up with "good" goals—specific, short-term, motivating, measurable.

A second approach is to write an imaginative scenario of your life some ten or so years in the future. Imagine you are writing a magazine article about yourself, and write a story you would be proud to see published. This written description will contain your long-term and medium-term goals, and suggest some practical steps to take now.

A third method is to scan the "laundry list" of goal areas in Figure 10–1 and write down the goal ideas triggered by this list.

Figure 10–1 A "laundry-list" of goal-setting areas

Family
  Relationship with spouse
  Relationship with children
  Household procedures
  Family activities
  Life style
  Other

Personal
  Spiritual
  Hobbies
  Recreation
  Adventure
  Learning
  Self-improvement
  Creative and cultural
  Other

Social and civic
  Community involvement
  Social service
  Friendships
  Social pursuits
  Clubs and organizations
  Other

Career-related
  Performance
  Responsibilities and positions
  Compensation
  Skills and knowledge
  Relations with boss, peers, subordinates
  Work choices, targets, and changes
  Professional involvement and visibility
  Other

Financial
  Income
  Savings
  Budget
  Investments
  Net worth
  Cash flow
  Other

Goal Types

The goals you defined in various goal areas (family, personal, etc.) will differ by type and time frame. Key goal types include maintenance, achievement, improvement, problem-solving, start/stop, and learning/development.

Maintenance goals are just what the name says—maintaining current performance at an acceptable level. Keeping your weight steady, continuing amateur photographic pursuits, and other stable pursuits fit this category.

Achievement goals entail specific accomplishments. Earning a desired promotion, solving a technical problem, directing an amateur play, and buying a new home are achievement goals. Once these goals are achieved, there is no further reason for them, except as platforms for reaching new ones.

Improvement goals involve doing better what you already do. Improving your tennis game, increasing reading speed, and developing your public speaking abilities are typical examples.

Problem-solving goals resolve troublesome situations which reduce your effectiveness or happiness. Such goals frequently concern improving difficult relationships or overcoming destructive personal habits. Examples include conquering procrastination habits, communicating better with a teenager, and patching up a damaged on-the-job relationship.

Start/increase/stop/decrease goals change the frequency of certain behaviors. You can start (or increase) satisfying and beneficial activities and stop (or decrease) those which are unfulfilling or detrimental. Start/increase goals may include spending more time with your family, reading more classical literature, exercising, and speaking up in staff meetings. Reducing drinking or smoking, cutting down on weekend work, and eliminating television sitcoms from your viewing regimen are typical stop/decrease goals.

Learning and development goals may be included above or stated separately. Continuous acquisiton of new skills, knowledge, attitudes, values, and behaviors is vital for both career progress and personal satisfaction. Examples include learning about data base management, taking a course in stock market analysis, expanding your vocabulary, deepening your spiritual beliefs, and so forth.

Your goals include long-range, medium-range, short-range, and immediate time horizons. Long-range goals are practical only if you can define short-range objectives which contribute to them. Similarly, immediate and short-range objectives should relate to important long-term goals. Ask how your goals relate in the time dimension. What short-range objectives would contribute to a long-range goal of becoming a respected authority in your field? What long-range goal does the short-range objective of learning about microcomputers help you achieve? Linking goals in different time frames strengthens your strategy.

ORGANIZE AND CLARIFY GOALS

Your goals will include a mix of different goal areas, types, and time frames.

Goal Trees

A good way to organize these diverse goals is by using "goal trees," illustrated in Figure 10–2. Goal trees visually relate long-term goals (at

Figure 10–2 Goal tree example

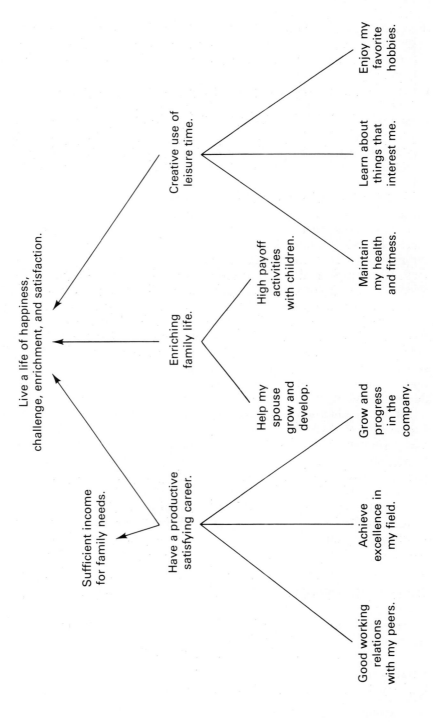

- Live a life of happiness, challenge, enrichment, and satisfaction.
- Creative use of leisure time.
    - Enjoy my favorite hobbies.
    - Learn about things that interest me.
    - Maintain my health and fitness.
- Enriching family life.
    - High payoff activities with children.
    - Help my spouse grow and develop.
- Sufficient income for family needs.
- Have a productive satisfying career.
    - Grow and progress in the company.
    - Achieve excellence in my field.
    - Good working relations with my peers.

the top) to the medium- and short-term objectives to achieve these goals (in the middle) to the specific actions necessary (at the bottom).

Looking up the tree explains *why* a particular goal is undertaken; looking down explains *how* it will be accomplished. The logical "if-then" relationship links goal levels. The logic inherent in goal trees states that *if* lower-level objectives are achieved, *then* the higher-level objective will be reached. With this logic, you can identify additional objectives required to strengthen the if-then linkage between goal levels.

Note that the logic in the diagram expresses your view of cause-and-effect, the relationship between *activities you can do* and *results that will be achieved*. Developing goal strategy means clarifying the linkages between goals at each level. Your strategies thus become specific, testable hypotheses.

Sketching goal trees makes an interesting paper-and-pencil exercise. Start by writing a goal, and below, identify the objectives required to reach the goal. For any goal, there are alternative combinations of lower-level objectives which can reach that goal. Each mix of alternatives constitutes a different strategy.

Identify Your Assumptions

As you consider various strategies, also identify the assumptions necessary for the linkages to be valid. Assumptions are factors over which you have little control, but which are necessary for cause-effect hypotheses to be valid. Assumptions include personal, interpersonal, organizational, and external conditions. If the assumptions are invalid, your strategy needs refining.

For example, if your supervisor plans to retire within a year, and you are interested in this job, your strategy may be to become the "obvious" replacement by demonstrating excellent current performance and acquiring new skills. Your hypotheses are:

If I perform well and build skills, *then* I am the superior candidate.

If I am the superior candidate, *then* I will be promoted.

You are implicitly assuming that your supervisor's job will be filled internally, that you are the superior candidate, and that the superior candidate will be selected. If those assumptions are uncertain, your strategy is shaky.

Examine all key assumptions and take steps to increase their validity. Reviewing past company promotion policy should indicate whether this job will be filled internally.

Reviewing the assumption that you are the superior candidate entails asking what skills the job requires and comparing your qualifications with others in the unit who may also want that job. This review will suggest what performance and skill-building areas to emphasize to become the best candidate.

Reviewing the assumption that the superior candidate will be selected requires examining how such decisions are actually made. If it turns out that superior candidates are selected and you are confident you are (or will be) that candidate, your strategy is sound as is. But if being known by so-and-so is necessary, your strategy will include getting known by so-and-so.

Every goal achievement strategy includes implicit assumptions. By making these assumptions explicit, you can examine each and adjust your strategy accordingly.

ANALYZE GOAL COSTS AND BENEFITS

Listing possible goals is not difficult if your imagination is working. The tough part is figuring out how to achieve them all. Every goal demands resources—time, energy, money—and these are limited. Thus you must review your many goal alternatives to identify priorities.

Not all your goal possibilities warrant serious attention. Some are in the "wish-list" category—desirable but not very realistic. Others could be reached, but only with more effort than you are willing to invest. For others you may lack motivation. Some are goals which other people have for you, which you do not want. As you review your goal alternatives, a critical few will stand out as desirable, realistic, and personally motivating.

There are multiple paths to any goal, some better than others. Conduct a rough "benefit-cost" analysis of the alternative routes to choose the best approach. You want to maximize the total goal payoff within time and personal resource constraints. The following "expected value" formula helps to evaluate goal and strategy alternatives and make choices:

$$E = \frac{P_s \times \text{Benefits of reaching goal}}{\text{Cost of goal achievement}}$$

where

$$Ev = \text{Expected value of goal payoff}$$
$$P_s = \text{Probability of success}$$
$$\text{Benefits} = \text{Perceived value of goal achievement}$$
$$\text{Cost} = \text{Time, money, and other resource requirements}$$

This formula helps to choose from alternatives even if you don't quantify all the factors. Consider an example:

Bill Wood always wanted to play the guitar for relaxation and enjoyment. He didn't expect to achieve concert level proficiency, but Wood wanted to become good enough to play in front of other people without embarrassment. One day he described his desires to a musician friend and asked how long it would take. The musician surmised that with a couple of hours of practice a day, Wood could play fairly well in four months. Wood mentally calculated the cost of the project to be 200 or so hours, plus at least $200 for the instrument and lessons. He mentally reviewed the status of his free time and bank account and immediately became discouraged. Learning to play the guitar cost too much for the perceived benefit.

The musician noticed his disappointment and suggested, "How about the harmonica? You can master it in ten hours and you can buy a harmonica and instruction book for ten bucks." Wood considered this option, and noted that the harmonica would also provide relaxation and enjoyment. The benefits would be somewhat less than playing the guitar, but at only one-twentieth the drain on his time and bank account! The benefit-cost ratio made sense; today Wood plays the harmonica with enthusiasm.

Every goal costs something. In addition, each goal has an opportunity cost—the value of alternatives not pursued. The opportunity cost to Wood of mastering the guitar was too high; he could gain more important benefits investing 200 hours in other pursuits. But the opportunity cost of learning the harmonica was reasonable.

Probability of goal achievement rests on the strength of the motivation. Research proves that personal motivation is greatest when achievement is reasonably probable—in the 30 to 70% range. Goals too easy, or too hard, are less motivating. Figure 10–3 summarizes the relation between probability of goal success and level of personal motivation.

DEVELOP GOAL ACHIEVEMENT PROJECTS

The best strategy for achieving goals is to organize them into *projects*. A project entails specific short-term objectives, an action plan, resource budget, and schedule. Well-conceived projects add leverage to your efforts and focus your efforts toward accomplishment.

Objectives

Objectives are "bite-size" short-term chunks of larger goals. Objectives appear at lower levels of your goal trees and, when achieved, contribute to your long-term goals. Select objectives which are:

Figure 10–3 Relation between personal motivation
and probability of goal achievement

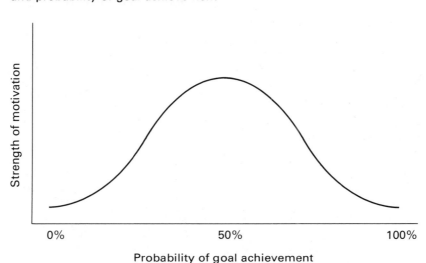

- *Motivating*   Something *you* want to accomplish, not what someone else wants for you. Unless you personally desire the objective,
  you will lack the commitment necessary for success.
- *Realistic*   Not so hard as to be impossible, nor so easy to require
  little or no effort. Keep them in the probability range that enhances
  motivation.
- *Specific, communicable, and measurable*   Defined so the result can
  be measured and discussed with someone else. Measurable objectives keep you focused and provide progress checkpoints along
  the way.
- *Result-oriented*   Stated as results to be achieved, not activities to
  be pursued. Activities are what you do to get there; results are
  what you find upon arrival. Add a completion date and a schedule
  that maintains interest and drive.

Figure 10–4 illustrates a typical project. An important long-term
goal has been divided into several short-term objectives. These, in turn,
are broken into specific, measurable activities to do each day or week.
Note the use of a monitoring system, with daily or weekly targets to
keep on track.

Aim for synergy when defining projects. Certain activities can
simultaneously provide multiple benefits and objectives. A daily jogging
program, for example, can improve your cardiovascular fitness, trim
your waistline, and build confidence. A project of visiting all local

museums and art galleries with your family is culturally beneficial, an enjoyable family activity, and low-cost, relaxing recreation.

Move Toward Goals Step by Step

The project concept enables you to accomplish long-term goals step by step. Consider the project undertaken by Mary Soares.

Soares was expert in meteorological application for agricultural planning. Her long-term goal was to establish her own consulting practice. Soares concluded that developing a national reputation as an expert in the field was necessary to reach this goal. She decided to write a book on the topic and set a medium-term goal of publishing one within three years.

Soares estimated that the medium-term goal, writing the book, would take 60 to 80 working days—a big task, with no certainty of getting it published. As a project, Soares decided to organize her notes, develop a book outline, and send proposals to possible publishers, a two-day task. Assuming favorable publisher response, she would undertake additional steps—collect her ideas, attend technical conferences, study the applications literature, and so forth.

The job of writing a book was massive and intimidating, but by breaking it into smaller projects, Soares' motivation to complete the effort increased. The key to achieving goals is to divide them into reasonably sized, personally motivating projects.

Monitor Project Progress

Visual feedback is a powerful method for tracking project progress. I find that simple graphs and charts add fun to goal achievement. In writing this book, I graphed the number of pages I wrote each week and made a game out of "beating" my targets.

To develop a monitoring tool, select your objective. Choose measurable behaviors and performance targets. Here are some measurable behaviors and targets for assorted objectives:

Improve my health and fitness

- Lose 10 pounds within 3 months
- Reduce my consumption of beer to less than 1 can per day
- Jog between 15 and 20 miles per week

Improve my leisure time quality

- Practice piano at least 4 times a week
- Reduce my time watching television to less than 2 hours per day on an average

Figure 10–4   Development project example

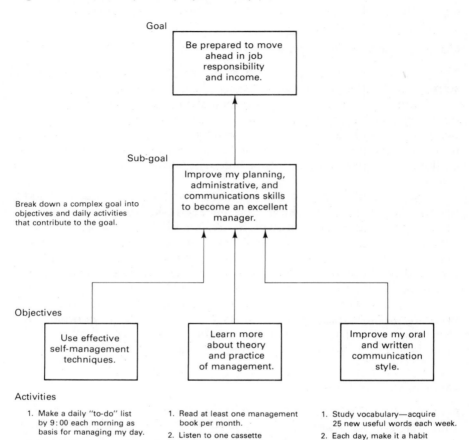

Goal

Be prepared to move
ahead in job
responsibility
and income.

Sub-goal

Break down a complex goal into
objectives and daily activities
that contribute to the goal.

Improve my planning,
administrative, and
communications skills
to become an excellent
manager.

Objectives

Use effective
self-management
techniques.

Learn more
about theory
and practice
of management.

Improve my oral
and written
communication
style.

Activities

1. Make a daily "to-do" list
   by 9:00 each morning as
   basis for managing my day.
2. Prepare (for my use only)
   a weekly written summary
   of my work on each Friday.

1. Read at least one management
   book per month.
2. Listen to one cassette
   learning program each week
   during my commute time.
3. Have one lunch per week with
   persons I consider good
   managers.

1. Study vocabulary—acquire
   25 new useful words each week.
2. Each day, make it a habit
   to speak with two people
   I don't know.

Improve my career skills

- Incorporate 50 new words into my vocabulary each month
- Meet 10 new people a month
- Invest 20 hours monthly in continuing education activities

Creativity is required to identify specific behaviors and select performance targets for some objectives. The objective "Develop better relations with my boss" does not easily plot. But you can choose measurable activities which contribute to this objective, such as "brief my boss on work status every week."

Figure 10–4 (cont.)

Goal Prepare to move ahead in job responsibility and income

Duration/date ___April___

Project	Achievement measure	2 M	3 T	4 W	5 T	6 F	7 S	8 S	9 M	10 T	11 W	12 T	13 F	14 S	15 S	16 M	17 T	18 W	19 T	20 F	21 S
Effective self-management																					
1. Make daily "to-do" list by 9:00 a.m.	Have list done at least 4 times a week.	✓	✓	✓	✓	✓	–	–	✓	no	✓	no	✓								
2. Prepare weekly work summary each Friday.	Summary complete and in my notebook.					✓							✓								
Learn about theory and practice of management																					
1. Read at least one management book monthly.	Complete "MBO" book in April.						2 chapters read					4 chapters read									
2. Listen to cassette learning programs.	One program per week.				✓	"Listening program"															
3. Lunch with good managers.	One lunch per week minimum.					✓ Fred			✓ Bill												
Improve my communications style																					
1. Study vocabulary.	Acquire 25 new words a week					30 words						45 words									
2. Speak with new people.	Talk with at least two per day.	2	1	2	4	3	2	0	1	4	2	0	3								

Figure 10–5 shows a monitoring method I use to keep physically fit. At the start of each month I set targets for various exercise categories—the number of situps, pushups, etc. I plot these as target lines and make an effort to stay above the target. I don't feel guilty about missing a couple days, as long as I meet my objective for the month. The graph keeps me motivated. I get psychic satisfaction from plotting my results each day, in addition to the physical benefits.

Be creative and devise ways to monitor your key objectives.

Figure 10–5   Project monitoring chart

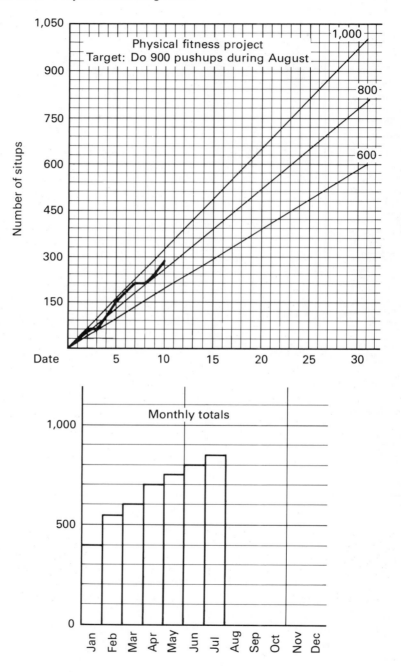

SUMMARY OF KEY POINTS

- Goal setting is fundamental to lifelong growth and vitality. Choose goals that are important to you in work, family, personal, financial, and social activities. Don't select the "ought to do's" unless they are also "want to do's."
- Write your goals. Putting them on paper helps you to be specific. Periodically review and revise your goals.
- Don't be overly ambitious. Start with a small number of attainable objectives. Gradually expand your efforts.
- There are alternative routes to every goal. Consider different approaches and select those which give you greatest payoff for your resource investments.
- Create goal projects. Break down big goals into specific, realistic objectives. Increase the success of your projects by keeping them moderately challenging—not too easy and not too hard.
- Make your objectives measurable, so you can review your progress along the way. Develop a chart or graph to monitor results.

QUESTIONS TO CONSIDER

1   What are your major long-term goals (five to ten years) in key goal categories? How would you prioritize them?

2   What are some of your medium-term goals (two to five years)? Short-term (less than one year)?

3   What are you willing to invest to achieve these goals? What are you willing to give up to reach these goals?

4   What are some projects you might undertake to reach these goals?

5   What are some things you'd like to start or stop doing now?

RECOMMENDED READINGS

Morrisey, George. *Getting Your Act Together: Goal Setting for Fun, Health, and Profit.* New York: John Wiley & Sons, Inc., 1980. A fun-to-read book using the Management by Objectives (MBO) process to identify, set, and achieve personal goals.

Kirn, Arthur G. and Kirn, Marie O'Donohue. *Life Work Planning.* New York: McGraw-Hill Book Co., 1978. A thought-provoking planning process developed by the Kirns from years of workshop experience. Useful checklists and planning forms. Greater emphasis on life than on career or work.

CHAPTER 11

Maintaining Success Throughout Life

The first forty years of life give us the text; the next thirty supply the commentary.

Shopenhauer

OVERVIEW

One's personal interests, values, goals, and concerns unfold over the years. Each person experiences life's different stages and transitions. You can enjoy satisfaction and success at each stage of life by recognizing the challenges each stage presents. This chapter examines:

- The interplay of career with family and personal concerns
- The key challenges of each career stage
- The predictable stages of adult development
- How to recognize change signals
- The multiple avenues for life enrichment at each stage

YOUR LIFE CONCERNS EVOLVE

Recently I found a dusty and forgotten diary I wrote ten years ago, and spent an afternoon browsing through my goals, aspirations, fears, and concerns of an earlier era. Some of those pages were so foreign I could not believe I wrote them; others would be written identically today.

Each person has a unique pattern of concerns, roles, values, and goals. These can be grouped into three primary categories—work, family, and self. Individuals define their own life structure by the unique emphasis put on each category and role.

130

The pattern changes over time. Work concerns may dominate the 25-year-old striving to make a professional mark. At age 35, now married and with children, the family dimension grows to equal importance. At age 55, with children grown and career interest declining, attention may shift to enriching personal concerns.

Figure 11–1 shows how the emphasis on different categories varies by individuals and with time. The lower part of the figure illustrates that the major categories are composed of multiple roles, plus new ones yet to be discovered.

Each role voices its own concerns and priorities. Some shout for attention, others speak quietly but persistently. Some are compatible, others in conflict. Conflicts force you to find a satisfactory balance among multiple concerns.

Figure 11–1   Individual life structures are unique

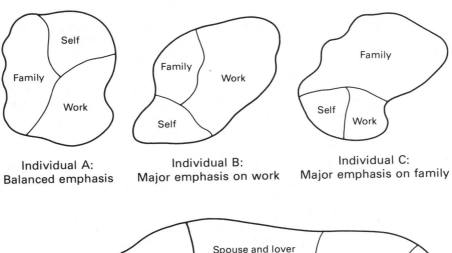

Individual A:
Balanced emphasis

Individual B:
Major emphasis on work

Individual C:
Major emphasis on family

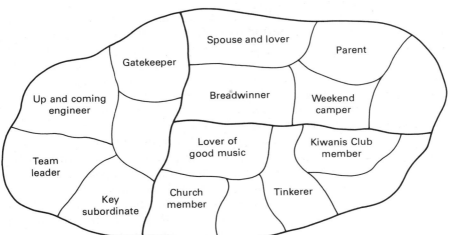

Role conflicts frequently appear as competition for limited time. We each have 24 hours a day, one day at a time, no more and no less. These 24 hours must be allocated so as to meet the key needs of yourself and important others in your life.

There never seems to be enough time. On the other hand, there are 24 hours a day, enough time for what is truly important. Henry David Thoreau once remarked that he didn't seek to find ways to save time, but sought better ways to use the time he had.

ANTICIPATE GROWTH AND CHANGE

Studies of adult growth have produced models which explain the adult development process. Keep in mind that social science models are different than those of the hard sciences. Models of human behavior lack the rigor and precision of scientific models. Social science models provide descriptive concepts for large groups of people but fail as predictive concepts for individuals. There are always personal variations on and exceptions to behavioral themes. We all have fingerprints, but each fingerprint is unique.

With this caveat, I'll briefly describe two models of special interest. They may help you understand your life experience to date and anticipate changes you may experience in the future.

Yale University psychologist Daniel Levinson studied the growth of adult males over a long period of time and documented his remarkable findings in *The Seasons of a Man's Life*. Levinson concluded that adult men go through a series of relatively smooth, stable periods of six to eight years, interspersed with rocky, transitional periods of four to five years.

Your major values and needs remain the same during the stable periods. But the smooth periods are inevitably followed by transitions, a time of self-questioning and re-evaluation. During these transitions, people make decisions concerning career, family and self which set the stage for the next stable period. These decisions can reaffirm the current course or involve minor or major change.

Another model (see Super et al. in the Recommended Readings) suggests that individuals progress through five age-related vocational stages. These five stages entail *growth* of interests and abilities (during childhood), *exploration* of work roles and acquisitions of skills (during adolescence), *establishment* of career capability and expertise (in young adulthood), *maintenance* of work contributions (during years of maturity), and *decline* of work involvement (in old age).

These five stages depicted in Figure 11–2 show an important point occurring at around age 45. This has become known as the mid-life crisis, an often painful period of personal appraisal from which you may emerge with continued growth and performance, maintenance of the present course but breaking no new ground, or stagnation and decline. I'll develop these points further in the rest of this chapter. Below, let's "listen" as individuals of different ages describe their career concerns.

Linda Carson, age 23, computer programmer, large life insurance company

"This is my first job since I received my master's in computer science and I'm not sure I like it. I always thought that working in a large prestigious company would be exciting and challenging. Well, I've been here six months and still haven't done anything more important than debug some old programs. My works seems so trivial. I have been trained to do much more— they're not using my skills.

"I anticipated a job where I would build some good friendships. People are pleasant enough, but not too personal or interested in me as an individual. I'm thinking about quitting if it doesn't get better real soon."

Carson is new to the world of work and facing adjustments typical of the new jobholder. Though formal education provides good technical

Figure 11–2 Stages of career development

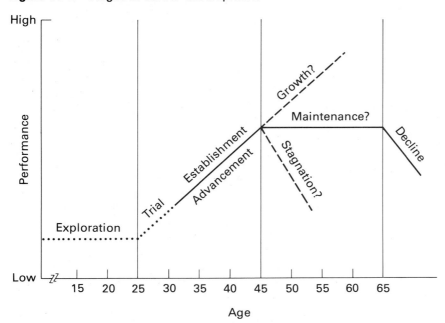

Adapted from *Careers in Organizations* by Douglas T. Hall.
Copyright © 1976. Scott, Foresman, and Company.
Reprinted by permission.

skills, it does not equip you for the task of learning how to work in the organization. Carson is low on the learning curve of work.

She is in the first career stage, apprenticeship, where you work on limited tasks under the close supervision of more senior professionals. This stage can be frustrating for new workers with a fresh head of knowledge who feel capable of doing much more.

The important learning Carson needs at this point is how to function effectively in an organization. She needs to acquire organization savvy and "learn the ropes," a task possible only through experience, not textbooks. She needs to discover her work personality and learn to influence peers and superiors to get things done.

Carson is in a testing period. She is building a reputation in the eyes of others which will influence how rapidly she earns more significant responsibilities. If she regards her tasks as trivial and doesn't carry them out efficiently, she'll miss the chance for better responsibilities. By performing well, she'll be noticed and find an organization mentor.

When management judges her as ready, she'll be assigned responsibility for larger portions of a project and work with greater independence.

Carson's job-change thinking is premature. She is in a high-demand occupation and can easily change, but still must climb the organization learning curve. Her present environment is as good as any for discovering her work style, studying other personalities, learning how things get done in a big company. Unless the job fit is a total mismatch, Carson would do well to stay. Too early and frequent job-hopping would earn her a reputation as an unstable worker.

Ron Simpson, age 32, mechanical engineer, industrial equipment manufacturing company

> "I earned a bachelor's and worked for two other companies before I landed here. I'm married with kids age eight and ten and have a nice house in a quiet suburban neighborhood a half-hour commute from work.
>
> "My company has a reputation as the best in the business. But I'm doing essentially the same thing I did five years ago when I joined. There isn't enough variety.
>
> "I've considered going back to school for an advanced degree. I need some way to get ahead of the financial game. Guys ten years ahead of me in this department aren't making much more than I am, and I don't want to get stuck salary wise. It seems the way to make it is to get into management—that's where the big bucks are, and that's why I'm thinking of getting an MBA. But that might be frustrating—pushing papers, administrative hassles. Besides, I prefer being near the equipment.
>
> "For the past four months we've been on a crash program which means lots of evening and weekends in the office. My kids play Little League baseball, and I've missed all of their games! That bothers me a lot. When I grew up, my father as always too busy with his work to spend time with

me. I don't want to make that same mistake, but I'm part of the company team, and I'm expected to sacrifice my personal time."

Ron Simpson is encountering what Levinson calls the "Age 30 Transition," a time of personal evaluation which typically occurs between the ages of 28 to 33. Simpson started his family early in life. Those who do so often find themselves wondering "What if I hadn't? What did I miss by commiting so early?" Those who defer marriage or serious work commitments during their 20s discover a need to settle down and build community and family roots.

Simpson gets considerable satisfaction from family activities, but his work schedule conflicts with those activities. The dilemma for professionals with children is that the years when children require the most attention are also the years when career imposes the greatest time demands. The only solution is to find a proper balance. Too many career-driven individuals realize, only after the children are grown, the loss of not spending enough time with them.

If the crash projects and overtime are normal practice in the company, Simpson's work-family conflict will grow and he will feel resentful. He should consider a change if his pattern is likely to continue indefinitely.

Simpson is experiencing both the effects of salary-compression and a lack of job variety. These are related issues he could address by taking a good look at his skills and interests, and examining where his company is going. His company examination should identify company growth areas or new projects he could become part of. His first stop should be the company personnel department, to get ideas from a career counselor and explore company-internal job change procedures.

If his talents are suited to marketing, he could likely increase salary there, or in a slot where he is more directly connected to sales or profits.

Downstown by Tim Downs

He could also explore smaller, faster-growing companies where he might find more work variety and chances for upward mobility.

Simpson mentions both his ambivalence about managing and his desire to get an MBA. Before committing to an MBA, he should try renegotiating his job to get more management experience and help him make the decision. He could take part-time courses directly relevant to his present job as well as move toward his MBA.

Simpson has taken the critical first step in making change decisions—willingness to listen to "inner data" and to clarify the reason for his dissatisfaction. An objective "where you are" analysis is the basis for redefining your life structure to achieve a more satisfactory balance.

Bill Perkins, age 43, optics research scientist, Department of the Navy

> "Something is gnawing at me, and I don't know what. Maybe it's the sense of growing older, and the feeling that I haven't contributed anything. If I could put my finger on it, maybe I'd feel better. I've tried several ways to solve the problem, but I found out that chasing young ladies wasn't the answer. Recently I started going to church again, to see if I can find what's missing. I'm also seeing a therapist.
>
> "I'm seriously thinking of shucking it all and moving back to the Midwest to run a hardware store or something like that. It would be less hectic than the city pace. But it would mean an income cut, and with two kids in college, I know that's not realistic.
>
> "I doubt I'll ever become a department head, which was my goal at one point. My performance here is good and I'm not concerned about losing my job. I can work here indefinitely—and that thought bothers me. I need the job; I like the people I work with. But I've lost the zip I once had."

Bill Perkins is experiencing what is called the "mid-life crisis," but the term is a misnomer. This key transition does not arrive at a precise mid-life date; the transition can occur anywhere between the ages of 35 and 50. And it need not be a crisis.

The key to navigating this transition and emerging with a deeper sense of personal renewal, vitality, and strength is to be prepared for the questions and challenges the situation presents.

Perkins is taking stock of his career and life to date, examining himself to develop a more satisfying structure for the future. He is having to adjust to the differences between his early career goals and the actual outcome. A key task at this stage is to accept the fact that earlier career goals may not be achieved. Those who achieve their goals face a different problem—their victories may seem hollow.

Perkins has not tried to supress uncomfortable feelings. He has stated what's bothering him—he won't become department head, he feels he hasn't contributed. He has also clarified some of his values, and is taking some construction action—going to church, getting professional guidance, and considering new career scenarios.

© 1981 Field Enterprises, Inc. Courtesy of Field Newspaper Syndicate

Perkins feels locked into his job. He is considering a move to the Midwest, perhaps to return to his roots, but is concerned about income drops. Unfortunately, Perkins has restricted his job change thinking only to second careers. By taking inventory of his transferable skills, and conducting a job search, Perkins could locate a challenging job in the Midwest that would *use* his skills and experience base.

He does not make clear why running a hardware store has appeal. He should think through that option and discover the attraction—is it the feeling of managing his own business? Or the folksy "helping people" environment of a country hardware? Or a fascination with gadgetry? By clarifying why the hardware store appeals to him, he can explore other possible options which satisfy this need.

Perkins can benefit by rebuilding his self-esteem. He has no doubt accomplished much during his work years. By recognizing his work and non-work strengths, accomplishments, and sources of satisfaction, he can consider other life-style options and may find ways to revitalize his current work.

A trusted friend who listens well can be a great asset in helping you define your feelings and clarify your options. Don't hesitate to seek professional help if you need it—it's not a cop-out. You visit the best doctor you know when your body is ailing; do the same for an aching spirit.

Bill Zealand, age 61, radar expert, telecommunications development firm

"I'm basically content with my career and life. Sure, I have my frustrations—everyone does. But I really can't complain; I love my work. My wife has her own career. We enjoy a decent life style. Our kids are grown and gone, and we are financially comfortable, though far from rich!

"I am fascinated by new radar development and I want to stay close to the technology 'til I retire. My company offered me a technical vice-president's job, but I turned it down.

"I've had a varied career. I started in industry, then taught college for a few years. Working with the kids was satisfying. But I got itchy and wanted something different, so I joined the Defense Department and played bureaucrat for a few years, managing a weapons acquisition program. After a few years it became boring, so I went to the private sector. Now I am assistant to the chief scientist, the best job in the company as far as I'm concerned. I can work on just about anything I choose. Our company does a first-class job, and I'm proud of the firm.

"I spend a lot of time with the new engineers—after all, they're the future of our company. I can tell which ones will be successful. The ones who will make it big are those with curiosity—curiosity enough to discover why an experiment didn't come out as expected. Those who don't care are already obsolete."

Zealand's comments prove that technical vitality doesn't have to decline with age. His enthusiasm demonstrates that obsolescence isn't a matter of age, but a state of mind, an attitude. He enjoys acting as mentor to the newcomers in his field. His career decisions show he is aware of his needs and skills, and has made appropriate choices—including refusing promotions.

Zealand is entering the "late adult transition" stage. During this stage he will make the shift from an active career to retirement (though he shows no indication he is about to slow down). This stage requires preparing to stop working. The transition to retirement can be abrupt and jarring, as a primary source of activity and identity is given up. Going fishing and relaxing quickly bore the person who had an active career.

There is no fixed time for retirement. It is more a question of energy levels, choice, and financial considerations. (The Navy's Admiral Rickover, for example, provided vigorous leadership well into his 80s.)

As he considers retirement, Zealand should also consider expanding his outside interests. He may want to make the work separation gradual, perhaps working part time or as a special consultant.

MAKE SMOOTH TRANSITIONS

How you confront the issues raised by life transitions determines the quality of the subsequent stage. People react to the problems by making decisions, many or few, large or small. Three outcomes are possible:

- *Maintenance*   Holding your own but breaking no new ground
- *Stagnation*   The burn-out syndrome
- *Continued growth and vitality*   Discovery of new interests and capabilities

Continued growth is within the capacity of everyone at every life stage. It begins by recognizing the "leading indicators" of life transition, including:

- Decline in energy levels, enthusiasm
- Lack of enthusiasm for life and work
- Loss of past interests
- Difficulty in sleeping
- Increasing illness; health problems
- Increasing drinking or use of drugs
- Irritability, moodiness
- A feeling of personal malaise

It's normal and *OK* to have these feelings. Refusing to accept and understand the problem signals, or seeking alcohol or drug escapes escalates the problem and complicates the solution. Recognize transitions as opportunities, and use your imagination to build more satisfying life structures. A 47-year-old industrial engineer confronting personal malaise created an opportunity which provided fulfillment:

> "I was in need of a new challenge, something to get me out of a mental rut. A year ago I found the ideal thing for me. I was talking with the owner of a local restaurant, and he described the problems they have clearing the parking lot when it snows. Well, I started thinking—if he has this need, lots of other people must also. I scanned the used equipment section of the paper and found a tractor and blade for sale, cheap. My brother-in-law was unemployed at the time, so I bought the tractor and put him to work. We distributed flyers to every house and small business in the neighborhood, offering quick service on snow removal. When the first snow hit, the response was tremendous. Since then, we've expanded into a year-round garden and maintenance service, doing rototilling and such in the summer. It's a real kick. I'm managing a real business, and it's growing as fast as we can buy more equipment. I'm tempted to take early retirement and go into it full time!"

Activities outside work are a rich source of fulfillment. There are dozens of enriching activities for personal stimulation. Your social needs can be satisfied by joining a civic organization, singing in the church choir, participating in recreational sports. Self-esteem can be enhanced by developing competency in a craft or hobby you enjoy. Self-actualization needs can be met by various learning, self-development, and cultural activities.

Experiencing lifelong enrichment requires recognizing the possibilities for growth and development at each life stage. Search creatively and take control of your opportunities for challenge and fulfillment now.

Max Ehrmann's famous essay "Desiderata" (Figure 11–3) eloquently captures a personal philosophy for maintaining success throughout your life.

Figure 11–3 Desiderata

"Go placidly amid the noise and the haste, and remember what peace there may be in silence. As far as possible without surrender be on good terms with all persons. Speak your truth quietly and clearly; and listen to others, even the dull and ignorant; they too have their story. Avoid loud and aggressive persons, they are vexations to the spirit. If you compare yourself with others, you may become vain and bitter; for always there will be greater and lesser persons than yourself. Enjoy your achievements as well as your plans. Keep interested in your own career, however humble; it is a real possession in the changing fortunes of time. Exercise caution in your business affairs; for the world is full of trickery. But let this not blind you to what virtue there is; many persons strive for high ideals; and everywhere life is full of heroism. Be yourself. Especially, do not feign affection. Neither be cynical about love; for in the face of all aridity and disenchantment it is as perennial as the grass. Take kindly the counsel of the years, gracefully surrendering the things of youth. Nurture strength of spirit to shield you in sudden misfortune. But do not distress yourself with imaginings. Many fears are born of fatigue and loneliness. Beyond a wholesome discipline, be gentle with yourself. You are a child of the universe, no less than the trees and the stars; you have a right to be here. And whether or not it is clear to you, no doubt the universe is unfolding as it should. Therefore be at peace with God, whatever you conceive Him to be. And whatever your labors and aspirations, in the noisy confusion of life keep peace with your soul. With all its sham, drudgery and broken dreams, it is still a beautiful world. Be cheerful. Strive to be happy."

SUMMARY OF KEY POINTS

- Strive for *balance* among all your life concerns and objectives. There are no easy ways to resolve the time conflicts between career and family/personal concerns. Be aware of your multiple needs and clarify tradeoffs to achieve a balance among your obligations.

- Multiple routes exist for life challenge and fulfillment; work cannot meet all your needs. Be creative, and identify fulfilling community involvement, social, and self-guided learning projects.

- Develop meaningful and satisfying extracurricular interests. These can provide stability in times of transition.

- Maintain frequent communication with at least one friend or colleague who is a good listener and whose opinions you trust. Test ideas, share anxieties, and assess options together.

- Welcome life transitions as an opportunity for growth. Note the signs that signal the need for change, and make decisions which increase personal growth and vitality.

QUESTIONS TO CONSIDER

1 What components of your life (work, self, family) are most important now? Within these broad categories, what specific elements are most important?

2 Five years ago, how would you have answered the preceding question?

3 How do you expect your emphasis on major work/life themes to change in the future?

4 What are the conflicts among your multiple life roles? How could you reduce the conflicts?

5 What are the areas of support and synergy? How can you increase this synergy?

6 What transitions have you encountered in the past? How did you resolve the rocky periods? What transitions are you experiencing now?

7 What are your major sources of stimulation and satisfaction now?

RECOMMENDED READINGS

Levinson, Daniel J. et al. *The Seasons of a Man's Life.* New York: Ballantine Books, Inc., 1978. An in-depth report from the team that researched the patterns of adult development. Provides a good understanding of the evolution of personal needs and concerns. Upbeat, with analysis of the possibilities for positive growth at each stage.

Greiff, Barrie S., M.D., and Munte, Preson K., M.D. *Tradeoffs: Executive, Family and Organizational Life.* New York: New American Library, 1980. An insightful look at the problems and approaches to achieving balance among work, home, and personal concerns. Written by two psychiatrists whose teaching and counseling methods at Harvard's Schools of Business and Law have won national renown.

Super, D., Crites, J., Hummel, R., Moser, H., Overstreet P., and Warnath, C. *Vocational Development: A Framework for Research.* New York: Teachers College Press, 1957. See pp. 40–41 for a discussion of vocational stages.

PART THREE

PUTTING YOUR PLANS INTO ACTION

CHAPTER 12

Finding the Job
That's Right for You

---

*Go, my sons, buy stout shoes, climb the
mountains, search . . . the deep recesses of the
earth . . . In this way and in no other will you
arrive at a knowledge of the nature and
properties of things.*
                              Severinus, 7th century

OVERVIEW

Career management means performing work that matches your skills,
interests, and career goals. For most of us, this boils down to periodic
job changes. Making effective changes requires a systematic approach
to identify, evaluate, and select from options. This chapter describes
how to:

- Determine when to make job changes
- Avoid the "Peter Principle" in your own career
- Use your contact network to identify opportunities in any field
- Find out what a job's all about before you apply
- Reduce the trauma of job layoff or termination

AVOID THE GRASS IS GREENER
SYNDROME

I periodically have lunch with my good friend Jim Bogaty. Our relationship is one of mutual personal admiration and good-natured job
"envy." Jim is an engineering manager for a large communications firm;
I am a self-employed management consultant.

As we discuss our work, we joke that we should switch jobs. We
both regard each other's job as desirable, but there are parts we don't
see.

He envies my ability to work for diverse clients but hasn't experienced the discouragement when I have few clients and little income. I envy the magnitude of the projects he works on but I don't appreciate the fact that his decision-making scope is limited. He envies my international travel; he doesn't appreciate the cost of overseas assignments on family life. I envy his pension plan and benefits but I don't take into consideration that these make him reluctant to switch companies. He envies my opportunity to write books but does not see the hundreds of lonely hours spent at the typewriter.

In reality, neither of us would be satisfied in the other job. Jim and I fall victim to the "grass is greener" syndrome. Neither of us fully appreciates our own job's advantages nor understands the drawbacks of each other's.

I think this syndrome strikes us all and leads to premature and too frequent job changing. There are some very good reasons for making changes, but all too many changes are made without a good reason.

Begin job-change thinking by understanding your motivation in making a change. This is crucial, especially if you are leaving a bad situation. Take careful stock of yourself, your job, your work environment. What is it that's lacking or bothersome? What are the conditions which, if present in your job, would cause you *not* to change? This helps to formulate criteria for your next job. A bad situation can be positive if it helps you specify what you are after.

Avoid job-hopping. Estimate how long it will take to fully master the job and maximize your learning. As a rule of thumb, this means two or three years if you're in your 20s, four or five years if you're in your 30s, and six to ten years in later life. This applies to changes between companies. Changes within the company can be more frequent.

The major factor in deciding when to change is the challenge and satisfaction the job provides. There is no need to change if the job grows as you do.

Your psychological contract includes assumptions about the job's ability to provide growth and satisfaction. These conditions change over time. The time to change jobs is before the growth cycle turns down. When your learning reaches a plateau, or when you can predict an end to growth in the future, start thinking of change.

If major changes are affecting the technology or business, accelerate your timetable. Don't be the last person to leave a sinking ship.

BEWARE THE PETER PRINCIPLE

The "Peter Principle," a term coined by Laurence J. Peter, suggests that a person advances to the point of his or her incompetency. The classic case of the Peter Principle is the top-notch engineer who fails as a manager.

This incompetency occurs because people are hired for their next jobs based on their performances in their current jobs. But the requirements of a new job may be totally different from what made someone successful in his or her old job.

You can eliminate poor job choices by thoroughly understanding what a job requires *before applying*. Too often people accept jobs that seem attractive, only to discover that the jobs demand skills they lack and don't use the skills they have. These mismatches are costly to the organization and the individual. Avoiding these mistakes requires a modified job-hunting approach, but the extra effort pays off.

CAST A BROAD NET

Unless you are absolutely certain what job you want next, don't prematurely narrow your alternatives. Job hunters typically limit their thinking to similar jobs and exclude potentially desirable alternatives in other fields, functions, and industries. A well-developed set of skills (coupled with awareness of skill transferability) provides you with extensive flexibility.

Cast your net broadly at the beginning of a search. Identify your choice criteria—the "must-haves" and "like-to-haves." Make sure your criteria include the key skills you want to use and personal or family considerations.

Also list general industry characteristics, nature of the work, the technology employed, and so forth. Think about your desired functions, the type of people you want to work with, the nature of the problems you want to tackle, and the skills you want to use. Add to this the characteristics of the organization (size, growth rate, specialty, etc.) and the type of position (research, project management, administrative, etc.). Then, and only then, narrow your choices down into possible companies and specific job targets.

USE JOB HUNT METHODS THAT WORK

The traditional way to hunt for a job is to have 50 or 100 résumés neatly printed, mail them out, and wait for responses. But the responses never come. This shotgun résumé approach doesn't work. One study concluded that it takes an average of 1,470 résumés to produce one job offer! Résumés are never hired; the best one can do is get you an interview.

How do people get jobs? A nationwide survey conducted by the U.S. Department of Labor concluded that:

- 48% of job applicants found jobs through friends or relatives.
- 24% found jobs by directly contacting employers.
- 13% used a combination of methods.
- 6% found jobs through school placement services.
- 5% got jobs through help-wanted ads.
- 3% were hired through public employment agencies.
- 1% found jobs through search firms and private employment agencies.

80% of all jobs are unadvertised, and these are generally the best jobs. They are filled by word-of-mouth methods. Advertised jobs are usually less attractive and produce a flood of résumés. To get the best, unadvertised jobs, you must penetrate the "hidden" job market.

TAP YOUR CONTACT NETWORK

Chapter 3 stressed developing a wide personal and professional contact network to penetrate the hidden job market. These people are your best resource for exploring job opportunities in virtually any field.

Assuming that you know about 250 people, and they each know 250 people, who in turn know 250 people, your "third generation" network of potential contacts exceeds fifteen million people! This network gives you access to virtually any career field.

Information Interviews

Earlier I discussed three categories of contacts, each valuable for different types of job change: your in-company contacts, for changes within the company; outside contacts in similar fields, for changes to other companies or industries; and friends and associates, for changes to totally new fields.

Your primary job-change objective is to make the best possible decision. To make a sound decision you must first thoroughly understand what your prospective jobs are all about.

The best source of such information are the current holders of that job. Use your contact network to locate these people and conduct "information interviews" with them. To illustrate the process, consider the most difficult change: to a field you know nothing about.

Finding Contacts

Let's say you are thinking about switching careers from microbiology to real estate. Begin by listing people you already know in the real estate field—friends, neighbors, members of clubs, churches, and other groups. If you identify people who are real estate agents or brokers, you are already on first base. If not, scan your list and look for people in *allied* fields who work with real estate agents—bankers, financiers, builders.

If you still come up short, identify people who are in "contact professions"—your insurance agent, pastor, and others who encounter persons of various backgrounds. Call them to explain that you are considering a career change to real estate, but you need more information to make a sound decision. Would they be kind enough to refer you to people they know in real estate, and would they also be kind enough to call these persons and mention that you will be calling?

Then, call these persons, introduce yourself, explain that you are considering a career change, and ask if they would be willing to spend a few minutes with you explaining their field.

Those who use this somewhat unorthodox approach report good results, for the simple reason that people like to talk about their work and enjoy being considered an "expert."

It may take a bit of courage to make the first call. But remember you need the best possible information when making an important career decision. Five out of six people you contact will be pleased to assist you. The worst that can happen is that they say *no*—not a terrible setback.

You only need one contact in the field to locate dozens of others, because you will ask each contact for the names of others in the field. By asking you might even locate former microbiologists who switched to real estate.

THE LOCKHORNS

The same principle works for less dramatic career changes and for job changes within the company. Let's look at how to conduct the information interview.

Conducting the Interview

The point warrants repeating: avoid career mistakes by thoroughly understanding the job requirements ahead of time. Apply for the job only after you know how you would spend a typical week, the main tasks you would encounter, and the skills and knowledge required to succeed. This concept holds true even for jobs similar to your current one and just down the hall.

The best way to develop this detailed understanding is by talking with people currently holding such jobs. The purpose of information interviews is not to *get* the job, but to *find out more about it*. Figure 12–1 lists some questions to ask. Add your own questions as well.

Don't limit your information search to one or two people: sample as many as possible. Viewpoints vary: different people describe the same job in different ways. With a variety of responses, you can compile a comprehensive list of the job's requirements. These interviews will give you the information to answer the following questions:

- Which of my key skills and abilities would this job let me use?
- Am I likely to be successful in this job? Challenged? Satisfied?
- What new opportunities will this open for me?
- What additional skills and abilities would I first have to acquire?

Now is the best time to conduct information interviews for future jobs of potential interest. Although you may not be planning a change

Figure 12–1   Questions for job information survey

How do you spend a "typical week" in this job?

What are your major responsibilities?

What aspects of the job are most challenging and rewarding?

What aspects are the most frustrating?

With what type of people (and for what purpose) do you work in the company? Outside?

What did you do before you held this job? Why did you take it?

What are the backgrounds of people who hold this job? (Look for special knowledge, education, and similar entry requirements.)

When people leave this job, why do they leave and where do they go?

In your opinion, what skills are needed to do the job well?

What special knowledge is needed to do the job well?

---

in the near future, you should have one or several possible next jobs in mind. You can prepare for future jobs now if you understand what your target jobs require.

Analyzing Skills Needed

The worksheet in Figure 12–2 is useful for analyzing how your skills fit with potential job requirements. From the knowledge gained in the interviews, list the key skills and specific knowledge required. Rate your current level in each. Your mentor or supervisor can help here by also rating you on these skills. Indicate the amount and type of development needed to acquire such skills.

If the gap between your skills and the target job's skill requirements is too great, forget about the job for now—there are other targets to consider.

Fill the gaps through your present job, supplemented by self-study as needed. Identify how your current job could be modified to give you this experience and negotiate these changes with your supervisor.

Applying for the Job

The best sources of information on available openings are through your contacts in the field and the individuals you interviewed. Call and ask about current or future opportunities in your area of interest. Ask about current openings, planned retirements, or expansions in their company and others. Even if you draw a blank, get the name and title of the person to contact should there be an opening.

If you have completed the earlier steps, you are ready to go for and land the job. The approach I recommend is a one- or two-page letter which summarizes how your skills and experience apply to this specific job. Don't include a résumé—draw from your résumé only those experience and skill highlights germane to the job of interest.

Letters are never hired, but a good one dramatically increases your chance of getting the interview. And because you have done your homework and understand the job, you'll come across knowledgeably. This topic deserves more attention than I have room for here; check the Recommended Readings at the end of this chapter for an excellent book on interviewing.

SURVIVE A LAYOFF OR FIRING

Not all job changes are voluntary. Getting fired or laid off can happen to anyone due to economic downturns, shifts in business strategy, technological change, reorganizations and mergers, not to mention chronic

Figure 12–2 Skills/knowledge analysis of target position
Position: _____

Key knowledge and skill requirements.	Check if area of strength.	Check if area of weakness.	If area of weakness, indicate the learning needed, personal interest/ability to acquire.	Activities which might help achieve desired level.

absenteeism, personality clashes, and poor performance. Whatever the reason, losing a job hurts.

Watch for Warning Signs

Few people see it coming. Keep your finger on the pulse of your organization and watch for leading indicators that termination may be in the works. Stay plugged into informal information channels. Be especially sensitive to information that your product's sales are slumping, the office budget is about to be cut, or the division is to be reorganized. Watch for changes in business direction that affect your unit, skills, or specialty.

Behavior shifts by your boss may signal an upcoming termination. If he or she suddenly becomes aloof, or if you receive an undeserved poor performance rating, gloom may be on the horizon. If you are no longer invited to key meetings or given new assignments, your days may be numbered. These are signals to get your résumé together and accelerate your thinking about job change.

If the signs are all there but you haven't been told, it's probably just a matter of time. If you have had a good working relationship in the past, you may want to approach your boss to discuss the situation. It may be advantageous to quietly resign if you think you can find a new job quickly and this doesn't jeopardize your unemployment benefits.

Dealing with Termination

How you handle the actual notification is critical. Try to maintain your control; don't scream or rant and rave. Be civil to your boss, even if you are mad as hell. You can be certain that potential employers will contact your boss before they hire you. If you can't change the decision, negotiate for benefits which will ease your transition to a new situation. Discuss these now or later when you've collected your thoughts.

Negotiate for the best possible termination terms. Try to remain on the payroll and keep the checks coming in long enough to find a new position. Bargain for a lump-sum severance payment. Get permission to use the office telephone and address while you are looking. Get your boss to agree to a plausible reason for the termination which doesn't reduce your credibility to prospective employers. Locate someone in the firm willing to give you a good reference.

Few people like to terminate others. Granting you reasonable termination conditions is one way supervisors can reduce their guilt feelings. If you don't get good terms, raise the issue at a higher level— companies are sensitive to their reputation and a shoddily terminated employee can sue, diminish worker morale, and bad-mouth them in the industry.

Understand the reasons for your termination. Is it because of business reasons beyond your control, or because of poor performance, inattention to the job, or other personal factors within your control?

Executive placement services and career counselors do a brisk business with terminated employees. But check these out carefully before signing up—most demand a hefty up-front payment and do little more than massage your bruised ego and help you write a résumé. Comparison shop: find out exactly what each will do for you, ask for the names of people in your position they have successfully placed, read the fine print and don't sign a contract on the spot. Employment agencies which list specific openings for which the employer pays the fees are a better bet.

The best defense against termination is a good offense. Maintain solid performance. Don't let personality differences get in the way. If you can't do that where you are, find a place where you can. Remain current—keep up with changes that affect your job. Stay active in your profession and build contacts which simplify job change. Monitor developments which indicate a possible downturn in your field or company. Termination is an unpleasant but survivable experience.

SUMMARY OF KEY POINTS

- Before you apply for a job, research it to assess the key skills required for success, a typical week in that job, and similar "what it's like" information.

- Your contact network can lead you to people holding jobs of interest to you. Don't be shy about using this network; most people are pleased to share the information you seek.

- Have in mind some functional objectives for your next job, even if you are not now considering a change. By being aware of what your target future jobs require, you can use your current work to help you prepare.

- Monitor the indicators in your industry and company which may indicate possible loss of your job. If it happens to you, keep your cool and negotiate for the best possible termination conditions.

QUESTIONS TO CONSIDER

1   What are some "next job" options you are considering? Can you summarize these options in a functional objective?

2 How do you define your "must-have" and "like-to-have" criteria for the next job?

3 What indicators will you use to know it is time for a change?

RECOMMENDED READINGS

Grundfest, Sandra, ed. *Peterson's Guide to Engineering, Science, and Computer Jobs*. Princeton, New Jersey: Peterson's Guides, Inc., 1982. A wealth of valuable information which includes employer profiles of 1,200 organizations, corporations, research laboratories, and government agencies. Breaks down employers by industry classification and size of work force. Over 700 pages, updated annually.

Medley, H. Anthony. *Sweaty Palms: The Neglected Art of Being Interviewed*. Belmont, California: Lifetime Learning Publications, 1978. A "how-to" book on handling yourself with poise during job interviews. Good tips on keeping control during interviews, overcoming objections, avoiding common pitfalls.

Gerberg, Robert Jameson. *The Professional Job Changing System: World's Fastest Way to Get a Better Job*. Parsippany, New Jersey: Performance Dynamics, Inc., Publishing Division, 1974. Authored by the founder of a leading outplacement consulting firm. Describes pros and cons of various job-hunt strategies and use of professional services. Good résumé examples and interviewing tips.

Coulson, Robert. *The Termination Handbook*. New York: Free Press, 1980. Useful advice whether you are on the giving or receiving end of terminations. Especially useful tips on negotiating for better severance terms.

CHAPTER 13

Getting the Most from Your Current Job

*The best place to succeed is where you are with
what you have.*

Charles M. Schwaub

OVERVIEW

The only way to reach your career goals is through your current job.
The future begins with the present, and by improving your current job
effectiveness, you solve present organization problems, build new skills,
increase job satisfaction, and expand your work options in and beyond
the company. This chapter examines how to:

- Increase your work productivity
- Analyze your job to identify needed changes
- Renegotiate your job to make it more challenging and satisfying
- Use your work as an R & D lab for personal growth

MULTIPLY YOUR JOB EFFECTIVENESS

Several time I have suggested that you work smart, not hard. Working
smart multiplies your effectiveness and performance. The key is to excel
in only the essential tasks. Not all your responsibilities deserve equal
attention. As Peter Drucker observed, "There is nothing so useless as
doing efficiently that which should not be done at all."

By concentrating on the essentials, and eliminating most of the rest,
you accomplish more each day and gain time by working productively.
By gaining just one hour per day, you create six additional work weeks
each year!

The 80–20 rule and "ALP" are two important concepts for increasing your job effectiveness.

The 80–20 Rule

The so-called 80–20 rule, developed by 19th-century economist Vilfredo Pareto, suggests that 80% of the payoff comes from only 20% of the activities, and the remaining 20% of the payoff comes from 80% of the activities.

The 80–20 rule on the job produces patterns like the following:

80% of your time is spent on 20% of your total activities.

80% of your interactions are with 20% of all the people you work with.

80% of the information in your file cabinets is devoted to 20% of the subjects you deal with.

80% of the payoff from a job comes from 20% of the activities you perform, and so forth.

Effective people clarify the critical 20% of the job tasks and concentrate on doing those well; ineffective people give equal attention and priority to all their job tasks. Focusing on the critical 20%, you can ignore or perform minimally much of the rest. But make sure that your view of the critical job elements is shared by your boss.

Each Task Has an "ALP"

Not every task demands equal performance. Essential tasks—the 20% that yield 80% of the value—must be done well. But the rest don't require excellent performance—a lesser level will suffice. The concept of the "ALP"—Acceptable Level of Performance—suggests that each task can be defined by the quality it requires.

The relationship between time and performance in Figure 13–1 holds true for most activities. The quality of the task increases as more

Figure 13–1   Identify the "ALP" (acceptable level of performance)

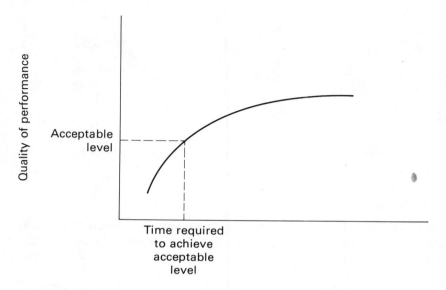

time is invested. But beyond a certain level, additional time investment yields diminishing performance returns.

Ineffective individuals fall into the "activity trap"—they let the available time determine the quality of their tasks. This tendency has been described by C. Northcote Parkinson in the principles "Activity expands to meet the time available" and "Tasks magnify in importance and complexity to fill available time."

Effective persons, by contrast, first define how good the performance must be. For each task, they ask what the ALP is and produce accordingly. For essential tasks (the critical 20%), they deliver more than the minimum; they deliver excellence. But for most tasks, an adequate quality will suffice. No real payoff occurs to you from doing excellently what can be done adequately.

Consider the project manager who writes a monthly status report that takes two days to prepare. Is that a reasonable level? It depends on how important that report is to the overall job. If the report is circulated (and *read*) company-wide, and if important decisions are based on the report, the answer is yes. An even greater time investment to rewrite the report and sharpen the key issues may be appropriate. But if it's a pro forma status report for the files, the writer may be able to get adequate quality with a one-day effort, and invest the additional day in higher payoff pursuits.

Scrutinize all your tasks to select the necessary results, the critical 20%, and concentrate on those. Subject all your work to the ALP concept. Define the quality of performance necessary and let that determine how much time you invest, not vice versa.

REBUILD YOUR JOB

You can lose the job enthusiasm you once had when you grow but the job doesn't. After you've mastered the job and face more of the same, boredom naturally results. Boredom leads to performance declines, discontent, and erosion of personal vitality. Some of the signals are:

General lack of interest in your work

Feelings of discouragement and malaise

Increased absenteeism, tardiness, daydreaming on the job

Decreased satisfaction from activities you formerly enjoyed

Irritation with your colleagues and boss

Decreased pleasure in your "nonwork" time

These signals are not problems, but opportunities to create or find a more stimulating work environment. You could find a different job elsewhere, but this is not always the best choice. You know the ropes where you are; your experience gives you important advantages.

Before seeking greener pastures, cultivate the one you are in. Try to "rebuild" your current job. Consider a job change only if the rebuilt job fails to satisfy.

You must take the initiative to redesign your job. It's your responsibility to engineer changes. The steps are to:

Analyze your needs to pinpoint sources of satisfaction and dissatisfaction.

Clarify organization unit functions, objectives, and opportunities to identify areas of potential challenge.

Identify change which, when made, would give you satisfaction.

Negotiate these changes with your supervisor.

Perform your redesigned job and continue to make changes as needed.

Analyze the Situation

Begin by identifying job aspects you wish to change. Your final objective is clear: to do more of what you do well and enjoy and less of what you don't do well and don't enjoy.

Be candid in clarifying the true source of your dissatisfaction. Often a "person-problem" is the cause, or a nonwork issue may need attention. If it's an interpersonal problem, treat it as such (the next chapter has some suggestions).

Figure 13–2 is a useful format to guide your analysis. Start by taking inventory of what you actually *do* on the job. Account for the time spent during a typical week, grouping your activities by similar functions or elements.

Next, estimate the percentage of time spent on the job function. Candidly rate how well you perform each function and how much you enjoy each. Also list functions you are *not* doing that you'd like to do. If management policies or procedures are the source of dissatisfaction, identify them.

Thus begins the redesign. The next step is to determine *how* to make these changes. This requires that you examine needs and opportunities in your work environment. Your chances of negotiating changes are increased if you redesign your work around objectives important to your boss and other key parties.

Clarify Organization Opportunities

Ask yourself the following question: What should my organization unit be doing, but is not now doing, which would challenge me and benefit the organization? Your answer will suggest opportunities for you.

Link your desired change areas to current or upcoming organization needs. Opportunities often come disguised as current *problems*. Your search should surface:

New projects, contracts, business areas

Upcoming task forces, study committees, and special project groups

New organization objectives and responsibilities for your work unit

Technological developments which affect the business

Successfully renegotiating your job depends on how your changes affect others and requires that the company benefit as well. Even companies with the most enlightened human resource policies will not let you "do your own thing" unless there is also payoff for them. But it's not an "either-or" situation of achieving your goals versus the organization's; it's finding a way to do *both*!

Identify Desirable Changes

The intent of job redesign is to match your needs and desires with the company's business objectives. The preceding steps should suggest change

Figure 13–2 Analyzing areas for job change

Key function or activity	Percentage of work time this involves	How well I do it Poorly Excellently	How much I enjoy it Hate it Love it
_____	_____	— — — — —	— — — — —
_____	_____	— — — — —	— — — — —
_____	_____	— — — — —	— — — — —
_____	_____	— — — — —	— — — — —

Organization or management policies and practices which create dissatisfaction:

Organization opportunities I would like to become part of:

areas; Donald Miller* suggests the following ways to incorporate these changes into a job redesign:

- Upstream and downstream shifting of duties to provide a greater sense of wholeness of activities and closure or completion.
- Temporary shifting of duties within a group to provide variety and new learning.
- Analysis of functions to identify those which are not rewarding and consideration of elimination or transfer of those functions.
- Increase in job loading by addition of new tasks or functions without elimination of old ones.
- Elimination of function or functions as unnecessary.
- Search for equipment or facilities which would when added change the nature of the work by taking over boring or repetitive activities or enhance human capability.
- Analysis of the proportions of different activities and gaining agreement for change in proportions.

*Reprinted from *Personal Vitality* by Donald B. Miller, copyright © 1977, by permission of Addison-Wesley Publishing Co., Reading, MA.

- Negotiations with management to change the goals of the group or organization in a way which will positively influence jobs.
- Negotiations with management to change the reward and punishment structure so that it more nearly matches the realities of the work and the needs of the group.

This list should give you ideas for specific changes. The next step is to redefine the "psychological contract"—revise the work agreement by negotiating changes with your supervisor.

Negotiate These Changes

Pick an appropriate time and setting—not when there are crises and everybody is fighting fires. Do your homework first and propose specific ideas.

Initiate the discussion with your supervisor on a positive, problem-solving note. You are not begging or pleading; you are providing creative suggestions for how you can achieve work unit objectives shared by your boss.

Explain how your ideas will benefit the organization. Demonstrate why your ideas make sense. If you recommend dropping current tasks you now perform, explain why those no longer are needed. If you are adding new ones, show why these are more important. Include in your recommendations some suggestions that may make your supervisor's job easier.

Focus on changes that can improve organization productivity. Suggestions on how to cut costs, enhance performance, and save time benefit the company and increase your value to the organization.

Don't expect automatic acceptance of even the best ideas; people are naturally reluctant to change. Your boss may need time to think about the implications of your suggestions for the rest of the work unit. If you are turned down listen carefully to the reasons and present a modified approach later.

An old expression says there is no limit to what you can accomplish if you don't care who gets the credit. Can you create a win-win situation in which you achieve your objective and the boss looks good in the organization's eyes?

CHANGE JOBS IN THE COMPANY

Many person-job matches just don't fit, and no amount of tinkering will help. If a job change is clearly needed, first consider other parts of the same company. You have inherent advantages in staying with the organization: First, it's what you know best, and a known devil is safer than an unknown angel. Second, you have transferable skills related to knowing how to operate in that environment or climate. Third, your change is not complicated by the external and life changes of relocation and family changes.

One way to rejuvenate a stalled career is to start fresh in an environment where people don't have preconceived notions of who and what you are. Lateral moves to other parts of the organization can mean exceptional growth and freedom. You have many internal mobility options, and organizations often encourage such changes. The strategies discussed in the last chapter apply to in-company as well as external changes.

USE THE ORGANIZATION AS AN R & D LAB

Your main responsibility is to perform the job you were hired for. But this doesn't mean keeping your nose to the grindstone and never looking up. Most professional jobs permit considerable discretion in how employees carry them out. Take advantage of this discretion.

View your organization as an R & D laboratory for acquiring new skills and knowledge to enhance your present effectiveness and prepare you for future jobs.

Modify the way you do your job to discover methods to make it more interesting. Take advantage of company-sponsored training programs.

Experiment with new techniques of interacting with other people. Your "people-experiments" may include testing ways to handle a domineering boss, speaking up in staff meetings, and contributing in leadership roles.

Don't confine your learning to acquiring technical knowledge. More important payoffs come from learning to influence people and organization systems.

SUMMARY OF KEY POINTS

- The best career development opportunity you have—and will ever have—is the job you hold now.
- Focus your best efforts on the few critical tasks which provide the biggest payoff—remember the 80–20 rule and the ALP concept.
- Work smart, not hard. Don't let activity expand to meet the time

available. Rather, figure out the level of performance required and invest only that level of resources.

- Virtually every job can be redesigned. To do this, analyze your current tasks, identify new opportunities, develop some specific ideas, and get agreement from your supervisor.
- To get organization acceptance of your desired changes, clarify the benefits to the organization.
- Use your job as a skills development laboratory and continually experiment with new ways of doing things.

QUESTIONS TO CONSIDER

1   What are the "critical few" tasks which give the greatest contribution to your job effectiveness?

2   What are the "trivial many" which consume your time but produce few payoffs?

3   How can you change your job to focus on the critical elements and reduce the less important?

4   What elements of your job would you like to expand? Decrease? Otherwise modify?

5   What upcoming opportunities give you the chance to restructure your job?

6   How can you link desired changes in your job with benefits to the organization?

RECOMMENDED READINGS

Lakein, Alan. *How to Get Control of Your Time and Your Life.* New York: Peter H. Wyden, Inc., 1973. The classic book for working smarter, not harder. Practical tips on priority-setting, scheduling, and getting important tasks done.

Knaus, William J. *Do It Now: How to Stop Procrastinating.* Englewood Cliffs, New Jersey: Prentice-Hall Inc., 1979. Helps you identify the behavioral blocks and overcome your excuses for delaying what you need to do. Good "procrastination inventory" checklist.

CHAPTER 14

Sharpening Your People Skills

The most important single ingredient in the formula of success is the knack of getting along with people.

Theodore Roosevelt

OVERVIEW

You cannot achieve career and life objectives alone; cooperation with and action by other people is always necessary. You can increase career power by improving your ability to communicate and work effectively with others. This chapter explores how to:

- Become aware of your impact on people
- Improve your ability to influence others
- Fine-tune your most important communication skills
- Apply these skills in job-related situations—meetings, supervising, negotiating, counseling others, etc.

Your day-to-day satisfaction at work *and* at home depends on the quality of your interpersonal relations.

PEOPLE SKILLS ARE CRUCIAL

Take a look around your company and you'll find that the most successful people, whatever their job function, have a well-developed set of "people" skills.

Career success demands good people skills. You cannot escape interpersonal relationships. Large, complex organizations reward people able to work successfully with people of diverse backgrounds, skills, and responsibility levels. Sharpening your interpersonal skills is vital; lacking these skills limits your career progress and decreases your satisfaction. Effective communication includes not only skills, but *attitudes* which foster feelings of openness and trust.

Mention "communications" and most people automatically think of speaking. However, the most important communications skill is not speaking but the critical skill of *listening*.

MONITOR YOUR LISTENING HABITS

Learn to listen—and listen well—and you master the most important skill of all. Most people like to talk; few like to listen. We learn the power of speech early in life. Your first squall from the crib brought a parent ready to feed or change you; grade school teachers praised you for the answers you gave, not your listening attention. We have been taught well to speak, but not to listen.

It's little wonder that most individuals in a two-person conversation wait impatiently for the other person to finish so they can speak and "score points." While waiting their turn, they rehearse their next idea, sharpen their argument, and do virtually everything but *listen*. "Communications problems" are the heart of most foul-ups. That's not surprising—the average listening retention rate of adults, 24 hours later, is a dismal 15-20%.

Listening Is More Than Hearing

Listening goes beyond *hearing*. Hearing is a *physical* act, listening a *mental* act demanding effort and attention. Your capacity to *process* information far exceeds the rate at which you *receive* information. You can absorb and process about 500 words per minute, but people normally speak only 150–250 words per minute. Thus there is excess capacity to daydream, mentally argue with the speaker, prepare the next comment—and miss the message.

The active listener uses this excess time to think about and try to understand the message. He or she listens both for *content*, the words in the message, and for *feeling*, the emotional context of the message. These don't always coincide. People frequently conceal their real emotions.

How to Be a Better Listener

Here are some tips to improve your listening.

1   Speak less, listen more. Limit your part of the conversation to less than 50% of the "air time."

2   Listen actively, not passively. Keep your mind on the subject. Use your excess mental processing capacity to summarize, evaluate, review, and truly understand the message.

3   Assume the other person has something interesting and important to say. If you make the opposite assumption, you won't listen well.

4   Listen patiently, even though you believe it may be irrelevant or wrong. Head nods or "mmm hmms" indicate that you understand, even if you don't agree.

5   Pay attention to nonverbal cues and body language. "Listen" for evasions and for what *isn't* said. These provide good clues to the emotional content of the message.

6   When the speaker addresses a point you want to know more about, simply repeat the statement in the form of a question. This encourages elaboration.

7   Demonstrate your interest in the speaker through eye contact. Maintain an "open" body position, leaning slightly foward in a relaxed position, arms unfolded.

8   Concentrate on the main message and ideas, not isolated facts. Take notes during or after the conversation.

By learning to listen well—to colleagues, boss, spouse, everyone—you encourage trust, respect, and develop personal influence. Semanticist and former U.S. Senator S.I. Hayakawa noted, "The funny thing about human beings is that we tend to respect the intelligence of, and eventually to like, those who listen attentively to our ideas even if they continue to disagree with us."

IMPROVE YOUR SPEAKING SKILLS

The reciprocal skill of speaking takes many forms—one-on-one discussions, problem solving in task groups, formal presentations to large groups. The ability to think on your feet, clearly express your ideas, and convince and persuade others has many obvious day-to-day work applications. Few people advance in their career without being effective oral communicators.

Increasing organizational complexity and size demands individuals able to "crosscut"—that is, work across department and function lines to solve complex technical problems. Persons who do this well are in short supply.

How to Be Better in Group Discussions

Here are some tips for improving your skills in one of the toughest communication situations—group problem solving. You can do the following whether or not you formally head the meeting:

1   Avoid esoteric vocabularies or special jargon which may be unfamiliar to others in the group; insist others do the same.
2   "Scope" your conversation. Introduce a complex or lengthy message with a general overview which helps others follow your message.
3   Encourage participation by silent participants who may have important points to contribute.
4   Observe group dynamics. Watch for "hidden agendas."
5   Periodically summarize and steer the conversation to the most useful objectives.
6   Speak no more than your share and listen intently. When you speak, you do not learn.

Develop the habit of speaking up in all meetings you attend. Many people with good ideas stay silent because of shyness. The group suffers from not having your ideas, and you are regarded as not having any. An example shows how one individual overcame shyness:

> "I was terrified to speak up or defend my ideas in groups because I don't think I am articulate. Then somebody gave me a great idea: practice speaking in nonthreatening situations. That made sense and I tried it. I initiated conversations with strangers wherever I went—waiting for the bus, in the line at McDonald's, wherever. Most of it was small talk, but I was surprised at how I could generate some interesting conversations. My speaking and listening skills improved and I gained confidence speaking out more at work. A colleague recently commented that he admired my style in groups!"

An excellent means to sharpen your communications skills is to join a Toastmasters Club, located in nearly every city and town (and within many organizations). Toastmasters provides a supportive, enjoyable atmosphere for fine-tuning your speaking and listening skills.

SHARPEN YOUR INTERPERSONAL INSIGHTS

Interpersonal effectiveness begins with self-awareness. In relating to others we attempt to meet our own needs for influence, respect, and control. But how about the needs of other people?

Effective communicators are aware of the ego and esteem needs of others. Do your interactions answer your needs at the expense of others? If so, you are manipulative. Do you consistently "cave in" to others? If so, you capitulate or withdraw. Or are your interactions mutually supportive, giving and taking, satisfying both parties? This is the style to aim for.

Developing your best possible communication style requires knowing how you affect others. We often "tune out" certain individuals we deem unimportant or uninteresting. Secretaries are a prime example of people we assume have little of importance to say. By ignoring or avoiding these people, we transmit unconscious signals that devalue them.

Respect Other People

To improve your communication style, regard each individual as important and unique. Demonstrate this through simple contacts—a smile, greeting, or acknowledgement. If your interactions make others feel good about themselves, they'll feel good about you. And you will be enriched for having enhanced others.

Do you unconsciously degrade or alienate others? Are you a "topper"—always trying to improve their story, example, or joke with your own? Or does your behavior build others up and make them feel good about themselves?

Get Feedback on Your Own Behavior

Unfortunately, we don't always know how others view our behavior. The "Johari window" in Figure 14–1 is a useful structural concept for explaining interpersonal behavior. Behaviors may be grouped into four quadrants according to whether they are known or unknown to ourself and to others.

These four combinations are respectively labeled open, hidden, blind, and unknown. Good communicators expand the open, shared category by *disclosing* information about self to reduce the "hidden" block and soliciting *feedback* to reduce what we are "blind" to.

American men with the "macho" image regard disclosure as a form of weakness. Be strong, don't show your feelings, they reason. They confuse disclosure with uninhibited, let-it-all-hang-out emotional gushing. To disclose means to honestly express your needs and ideas. This is not weakness, but personal strength.

To eliminate your "blind" spots—what is unknown to you but known to others—be sensitive to how others regard you; even ask to get the feedback you need. Ask your mentor or a close personal friend how he or she thinks you come across.

Figure 14–1   Johari window

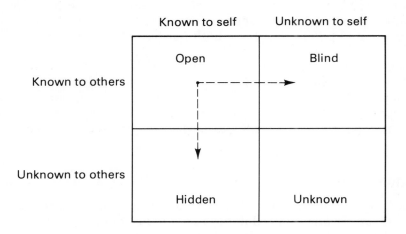

Source: Originally developed by Joe Luft and Harry Ingham.

GET AND GIVE PERSONAL FEEDBACK

You can expand your self-perceptions and reduce your blind spots through feedback from others. Mentors are extremely useful for this purpose, as are friends and colleagues who know you well. Your boss is an ideal resource if you enjoy a relationship of mutual trust and respect.

The feedback you receive may be disconcerting, but keep in mind that your objective is to get data to expand your awareness of self. The choice of whether to act on this data is up to you. Initiate your request for feedback in a straightforward, non-threatening manner:

> "Charlie, you've known me and my work for a long time, and I respect your observations. I'm doing some thinking about my career strengths and weaknesses, and your observations would be extremely useful to me. As you see it, what are some of the things I do well, and what are some areas that I might work on?"

Make it easy for the other person to provide observations. Listen without arguing, disagreeing, or defending. Just listen. Above all, be cool and in control of yourself. If you respond to something you don't like with anger or physical abuse, you won't get useful feedback from this source in the future (you may even get arrested!).

You can help others grow by providing them with feedback. The most useful feedback is *solicited* rather than *imposed*—the other person should be open and receptive to it.

The most useful feedback describes only your reactions to the person's behavior; it does not evaluate the behavior. Leave evaluation to the other person; if you do it, he or she will get defensive.

Be as specific as possible. Describe specific incidents that lead to your reactions, rather than providing general comments.

Direct your comments towards behavior which the person can control. When the other person is reminded of shortcomings that can't be changed, frustration results. Make the process useful to the other person. Don't use this to "dump" so that you feel better.

PREVENTING MANIPULATION BY OTHERS

Good communications occur when both parties share constructive ideas and attitudes. But when others use manipulative power tactics, you

Figure 14–2 A credo for my relationships

You and I are in a relationship which I value and want to keep. Yet each of us is a separate person with unique needs and the right to meet those needs.

When you are having problems meeting your needs, I will try to listen with genuine acceptance, in order to facilitate your finding your own solutions instead of depending on mine. I also will try to respect your right to choose your own beliefs and develop your own values, different though they may be from mine.

However, when your behavior interferes with what I must do to get my own needs met, I will tell you openly and honestly how your behavior affects me, trusting that you respect my needs and feelings enough to try to change the behavior that is unacceptable to me. Also, whenever some behavior of mine is unacceptable to you, I hope you will tell me openly and honestly so I can try to change my behavior.

At those times when we find that either of us cannot change to meet the other's needs, let us acknowledge that we have a conflict and commit ourselves to resolve each such conflict without either of us resorting to the use of power or authority to win at the expense of the other's losing. I respect your needs, but I also must respect my own. So let us always strive to search for a solution that will be acceptable to both of us. Your needs will be met, and so will mine— neither will lose, both will win.

In this way, you can continue to develop as a person through satisfying your needs, and so can I. Thus, ours can be a healthy relationship in which both of us can strive to become what we are capable of being. And we can continue to relate to each other with mutual respect, love, and peace.

Reprinted with permission of PEI Books, Inc. From *Leader Effectiveness Training* by Dr. Thomas Gordon. Copyright © 1977 by Dr. Thomas Gordon.

must counter their attempts with special assertiveness techniques. Here are three useful tools.

Fogging

The "fogging" technique neturalizes attacks against you based on criticism of your past actions. If a past mistake of yours is used to justify a current criticism, don't react by arguing or defending yourself against such charges—this steers the conversation in a non-productive direction. Simply *acknowledge* the criticism and move on to your objective. Don't resist; neutralize the comment by letting it get lost in a "fog bank."

For example, assume you want to be part of "project Y," a "hot" upcoming project, and your boss counters by recalling your earlier performance on Project X. The dialogue might proceed as follows:

> You: "I'd like to be part of the 'Project Y' team."
>
> Boss: "What? After your past performance on 'Project X?' You were late on almost every deliverable."
>
> You: "You're right. I wasn't always as prompt as I should have been. As you know, that was my first project, and I've gotten a much better handle on how to deliver on time. My current experience provides the technical perspectives that should be on the team— don't you agree?"

Note the pattern. Rather than arguing with your boss's comment, explaining why you were late, or blaming it on others, you simply acknowledge the statement and move on to make your point, adding some positive reasons to your argument.

Broken Record

If "fogging" fails, you need to counter with heavier artillery, specifically the broken record.

This tactic is useful when others ignore or avoid the concerns you raise. You repeat like a "broken record" what you want, until the other person addresses the issue. An example:

> You: "The Acme project involves much more data analysis than we anticipated. I'd like two analysts put on the project to get it finished by Friday."
>
> Boss: "I knew we should never have taken that project! It's been a pain in the ass ever since we started."
>
> You: "I need two analysts to help with the data analysis."
>
> Boss: "I tried to tell them upstairs not to take the job, but they didn't listen. Now you're telling me we can't get it done by Friday."
>
> You: "I need two analysts to help work on the project."

> Boss: "If I were managing this company, I'd refuse a project like Acme. We can't make any money on that kind of deal anyway."
>
> You: "I need two analysts. How about Brock and Pottinger?"
>
> Boss: "Brock and Pottinger? Well, okay, I guess we have to get the damn thing finished."

If you keep returning to your point, the other person must eventually respond.

Diplomatic Disagreement and Confrontation

The art of diplomacy lets you disagree with and confront other people without creating an argument. This technique calls attention to contradictions, rationalizations, misinterpretations, or differences. Such differences can be between what people *say* they do and how they really *behave,* or between statement reality.

The idea is to acknowledge the other person's position ("you said ... but look") and add a suggestion or additional information. As an example, assume you are trying to get approval to attend a one-week management seminar offered by the training department.

> Boss: "I'm afraid I'll have to say no to your request. The office is pretty busy right now."
>
> You: "I know that it's busy—it's always busy. But I'm caught up on my tasks. I think the seminar will give me some good ideas for doing things more efficiently around here."
>
> Boss: "Well, I'm not sure that the seminar is going to do you any good."
>
> You: "You once said that you want us all to develop our potential. The seminar subject would help me do just that. Did you know that I'm the only person in the unit who hasn't attended a company training program in the last year?"
>
> Boss: "Well, maybe you're right. Let me think about it awhile."
>
> You: "I'm pleased that you see value in my attending the program. Why don't I go ahead and complete the training request form. I'll bring it to you for your signature tomorrow."

Don't let this turn into an argument, or attack the other person's ego— that is destructive. But be diplomatically persistent in pursuing your point of view.

SUMMARY OF KEY POINTS

- Become the best communicator you possibly can. Good "people-persons" enjoy more job options and get more from their work.

- Practice your communication skills in low-risk situations, on and off the job. Overcome shyness by initiating conversations with strangers.
- Study the communications style of those you consider most effective. Learn what you can, and try some of their tactics on your own.
- Put these concepts to work off the job as well. The art of active listening works wonders with a spouse or teenager.
- Enroll in a community college or local workshop to sharpen your skills and get insights into your own style. The Toastmasters Club is an excellent public speaking program.
- Virtually every situation you are in provides the chance to use your communications skills.

QUESTIONS TO CONSIDER

1   What communications styles do the most effective people in your organization adopt?
2   How do you think others would describe your interpersonal style?
3   What is the strongest aspect of your communications style?
4   What is the weakest aspect of your communications style?
5   Who are some people you might ask for feedback and constructive criticism?

RECOMMENDED READINGS

Cannie, Joan Koob. *Take Charge: Success Tactics for Business and Life.* Englewood Cliffs, New Jersey: Prentice-Hall, Inc., 1980. Good ideas for understanding your interpersonal style. Extensive treatment of interpersonal communications techniques.

Cohen, Herb. *You Can Negotiate Anything.* Secaucus, New Jersey: Lyle Stuart Inc., 1980. Sensible ideas and tips about dealing with your boss, banker, friends, children, spouse, and self. Especially good discussion on overcoming manipulative tactics by others. Well-written, witty, and conversational in tone.

CHAPTER 15

Learning for Your
Job and for Yourself

---

*The test and the use of man's education is that
he finds pleasure in the exercise of his mind.*

Jacques Barzun

OVERVIEW

Every effective career strategy includes lifetime learning. To rely on
yesterday's knowledge in a rapidly changing world erodes personal vitality
and invites career disaster.

Lifetime learning goes beyond career-related skills acquisition:
learning for pleasure, growth, and personal enrichment is equally vital.

This chapter presents tips to guide adult learners in how to:

- Define your learning objectives
- Discover how you learn best
- Achieve balance in your learning plan
- Better use your limited learning time
- Develop a learning contract

OBSOLESCENCE IS A PERSONAL CHOICE

Enough has been written about the perils of technical obsolescence to
frighten many competent people into believing they'll someday wake
up stricken by a dread disease that incapacitates their learning faculties.
To be sure, the explosion of technical knowledge demands a special
effort to keep current, but this task is within the ability of every individual. It's not the gray matter that erodes, it's the desire.

Obsolescence is self-chosen. It happens when you think you know enough to stop learning. Obsolete individuals mentally say the hell with it all. They protect their personal niches and let the rest of the organization fend for itself. They do the minimum the job requires and are present in body, but not in spirit. Personal and professional obsolescence coincide; lack of learning erodes personal vitality and leads to performance decline.

The career risks of letting yourself obsolesce are obvious. Plenty of competent people are eager to replace the underperformer. Organizations cannot afford to keep non-producers on the payroll. They may not be fired outright, but simply bumped aside. Lifetime learning is not a luxury; it is integral to every career and life strategy.

While the content of your learning program is unique, your learning objectives fall into four categories:

Professional development in your current job/field

Professional development in future areas of interest

Personal development and skills improvement

Interest expansion

The topic of lifetime learning can bring chills to the spine of those who recall their college learning experience. I still have occasional nightmares about final exam week. The repeating dream finds me months behind in my studies and unable to recall what courses I am even registered for. I wake up screaming, sweating, and thrashing the blankets; an understanding wife soothes by reassuring me I'm now in the "real world."

Fortunately, adult learning is different. Sitting at the feet of wise professors is minimal for the professional's learning strategy. I'll give some tips for using your adult learning strengths shortly; first here are some issues to consider for developing your lifetime learning program.

DEVELOP YOUR CAREER LEARNING PROGRAM

Develop your career learning program by examining issues concerning your career strategy and objectives, and issues concerning your organization's strategy and objectives. The following questions should help shape your learning program.

How is Your Industry/Profession Changing?

Gear your learning to the changes affecting your industry and profession. Chapter 2 suggested some ways to examine these.

If you are in a rapidly changing technology you need no reminder that the profession is changing. How do these changes occur? Are they evolutionary and incremental, or do new developments occur as step functions which make obsolete most knowledge acquired earlier? Capable engineers on the Navy's Polaris missile program found their skills dated for the Poseidon program; the Trident program was built on even newer technology. To update their knowledge was not impossible, but it required discipline and special study.

Draw on the thinking of your colleagues, professional bodies and technical associations to define the learning implications of change.

Where is Your Company Going?

How will your company change? What will it look like five years from now in terms of major products, key technologies, and growth areas? What employee skills will be in greatest demand? Those plans are now on the company drawing board, in the form of strategic plans and future personnel forecasts.

Get hold of your company's future personnel forecasts. Human resources departments generally project their future work force requirements three to five years ahead by job and skill category. Identify "hot" growth areas you might become part of.

This strategy makes sense even if you're not sure you'll remain with the company that long. Other firms in the industry will experience skill demands similar to your company's. By concentrating on acquiring new skills and knowledge fundamental to your company's future, you multiply your job options throughout the industry.

How is Your Job Changing?

The job you now hold is the starting point of your career learning program. Scrutinize your job tasks and responsibilities, using the 80–20 concept to clarify those most essential. (See Chapter 13.) What tasks should you do *exceptionally* well—the critical 20%—and what new skill acquisition does this imply? How will the job change in the next couple of years? What new projects will your office undertake? Consult written plans and other people if you are uncertain.

Two questions are especially useful: First, what is it that if you *don't* learn, will leave you behind? Include this knowledge in your program as a *defensive* strategy. For instance, even though your job may not directly involve computers, it is obvious that familiarity with automation is essential for future career progress.

Second, what is it that if you learn, will put you ahead? Engineers who learn matrix management methods acquire a high-demand skill which makes them valuable on task forces and special projects. Look for ways to broaden your skills package to create a distinctive career edge.

The life-cycle concept of Chapter 2 applies to career skills as well. The demand for skills in the industrial marketplace evolves as the industry faces new challenges. Concentrate on building high-demand skills in emerging and growth stages.

Where Are You Going in Your Career?

What career future are you contemplating? Do you plan to leave, or to remain with, your specialty, job function, profession, or industry in the next five years?

If you are planning to change, address your learning to the new areas of interest. If it's an industry change, learn more about industry structure and problems. If it's a change in your function or specialty, use the earlier tips for analyzing skill requirements to build skills for future jobs you are considering. If you are considering changing professions, focus on building transferable skills as discussed in Chapter 7.

If you plan to remain, concentrate on acquiring the knowledge and skills required to be effective in solving future problems in your specialty, function, profession, and industry. However you decide, distinguish between skills and knowledge for you to:

- *Acquire* Enhance or increase your capabilities
- *Maintain* Retain your current capability level
- *Neglect* Permit your current capability to decline

How Much Do You Need to Learn?

You can spend six hours or six years studying nuclear physics. In both cases you "learn" about the subject, but with some obvious differences in mastery.

How much to learn and when to stop depends on how you plan to use the knowledge. Do you need a "can do" competence level or will a "can talk about" level suffice?

Suppose you are a unit manager who must make decisions about purchasing a small computer system. You don't have to become a systems analyst, but you do need sufficient knowledge to make an intelligent decision. You would need to be able to:

Determine when it is appropriate to use a mini- or microcomputer

Define cost-effective applications of this equipment

Prepare technical specifications used to procure mini/micro equipment

Evaluate and select a mini- or microcomputer from the equipment on the market.

With learning needs defined, you can examine various alternatives for acquiring this knowledge. To decide when you have learned enough, clarify what you plan to do when such knowledge is acquired.

What Learning Resources are Available to You?

Much of your learning will come informally on the job, through new assignments, reading the literature, interacting with colleagues, and so forth. Take advantage of formal company training programs to acquire skills of immediate or future value.

Beyond this, investigate local college programs, books, libraries, universities, and self-study opportunities. Most communities of size have "open universities," with informal courses conducted by individuals in their home or a public place. These are good for interest expansion and sampling a new subject area before tackling it in-depth.

If you subscribe to periodicals and journals whose publishers rent their subscription lists, you probably receive a flood of brochures advertising two or three day intensive seminars. These seminars, conducted by top-notch professionals, are ideal for learning the state of the art.

What Are You Interested in Learning?

Your job/industry/company analysis will surface many learning opportunities, but you must also be interested in learning about these topics. If you lack interest, you'll lack the motivation to follow through.

Behind a lack of interest may be a misunderstanding of the topic or a fear you're not capable of learning. Before you make this conclusion, find out enough about the subject to make a valid decision.

For years I assumed that decision-tree analysis required extensive study and was of limited use to me. One day while waiting in someone's office, I found an easy book on the topic. Thumbing through the book sufficiently piqued my interest to learn more later.

How Much of Your Own Resources Should You Invest?

Most companies underwrite the cost of formal learning programs. But company policies vary, and you may want to take courses in subjects not covered by the company.

Be willing to invest your own financial resources in continuing education. An example shows how this can pay off:

"It was obvious that sooner or later our company would have to automate their word processing system. The typing backlog and secretarial shortage was costing us. We hadn't moved beyond the electric typewriter era.

"By chance, I saw an advertisement in the business section of the newspaper for an upcoming seminar on office automation. I was intrigued by office of the future topics. The company recently sent me to training and I didn't think the company would pay, so I took three days of vacation and invested $400 of my own money to attend.

"I returned to the office bubbling with ideas on how we could catch up with the rest of the world. I described the seminar to my boss and he mentioned that the company was forming a task force to study word processing options. I contacted the person in charge and ended up on a task force with members from each department to study options and make recommendations.

"I was the in-house expert on a topic I knew nothing about two weeks earlier. As it turned out, I became the de facto project manager, and we conducted a first-rate study. In the process I worked with all parts of the company and my stock rose incredibly. For me, this personal R & D investment really paid off."

How Do You Learn Best?

Research by psychologist David Kolb has shown that adults use four different learning styles to acquire new skills.

- *Concrete experiences* Learning through day-to-day jobs and by working with more experienced superiors and co-workers
- *Reflective observations* Learning by "standing back" and carefully observing a work situation
- *Abstract conceptualization* Learning through abstract ideas, building models, reading, and formulating theories.
- *Active experimentation* Learning by experimenting with approaches until they find methods that work best for them.

Most of us use all four styles at different times and with different problems. These four learning styles form a "learning cycle," as shown in Figure 15–1.

To be most effective, use all four styles. Involve yourself in new experiences (concrete experience), observe and reflect on these experiences from different perspectives (reflective observation), develop concepts to integrate these observations into logically sound theories (abstract conceptualization), and use these theories to make decisions and solve problems (active experimentation).

The best learning method is one I call "eclectic outreach." That is, don't confine your learning to a single method; use a variety of resources. Your colleagues are a rich source of information on emerging trends. Identify the "information gatekeepers" in your organization, those peo-

Figure 15–1   The adult learning cycle

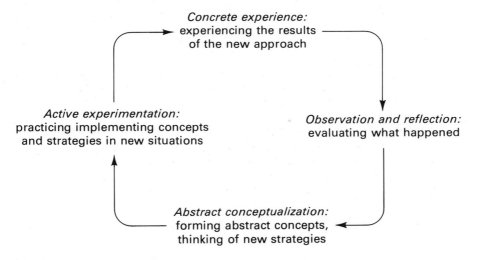

Concrete experience:
experiencing the results
of the new approach

Active experimentation:
practicing implementing concepts
and strategies in new situations

Observation and reflection:
evaluating what happened

Abstract conceptualization:
forming abstract concepts,
thinking of new strategies

Kolb, David A., Rubin, Irwin M., and McIntyre, James M.
Organizational Psychology: A Book of Readings. 2nd ed,
© 1974, page 28. Adapted by permission of Prentice-Hall, Inc.,
Englewood Cliffs, New Jersey.

ple who screen, synthesize, and stay abreast of the most important new developments. Encourage those in your work unit or carpool to share their specialized knowledge. Find out how the most successful people you know manage *their* learning. Talk with people in other departments and fields about what's happening in their areas. Join professional societies, and attend their meetings. Scan journals and magazines; maintain a clipping file. Enroll in seminars, company, and outside programs.

Don't confine your learning and experimentation to acquiring technical knowledge. More important payoffs come from learning better ways to interact with and influence other people and the organization system.

HOW ADULTS LEARN

As an adult, you learn differently than you did during your school years. Adult learning does not mean memorizing facts and formulas to regurgitate in final exams. Cyril O. Houle* studied adult learners; his conclusions have merit.

*From *Continuing Your Education* by Cyril O. Houle. Copyright ©
1964, McGraw-Hill Book Company. Used with the permission of
McGraw-Hill Book Company.

1   Act as though you are certain to learn. Nothing so disturbs begin-
    ning adult students as the nagging fear that they will not be able
    to learn what they would like to learn. Nothing is more reassuring
    than the discovery through experience that they can succeed. Adults
    can learn most things better than children, though it may take them
    longer to do so.

2   Set realistic goals—and measure their accomplishment. One fre-
    quent obstacle to adult learning is that men and women, realizing
    that they have the full power of their strength and vigor, think that
    they ought to be able to learn without any effort or strain whatever.
    In any learning program, therefore, you must first of all be realistic
    about what you can achieve.

3   Remember the strength of your own point of view. Your learning
    is strongly influenced by the point of view you bring to it. . . . Most
    important of all, do not let your established values harden into such
    fixed beliefs that you cannot tolerate new ideas. When this happens,
    the process of education ceases.

4   Actively fit new ideas and new facts into context. Your greatest
    asset as an adult learner is the fact that your experience enables
    you to see relationships. When a new idea or fact is presented, you
    can understand it because you have background and perspective.
    And you can remember it because you can associate it with what
    you already know and therefore give it meaning.

5   Seek help and support when you need it. Sometimes adults will
    choose to learn alone, and sometimes they will choose to learn with
    others. A balanced learning program combines many elements,
    though not all at the same time. But while adults often teach them-
    selves what they want to know, they may run into real dangers if
    they rely on this method too consistently. . . . One time when it is
    well to seek out a teacher is when you are beginning a study of a
    new subject. . . . A second time when you need help is when you
    bog down in your studies. . . . A third time when it is wise to seek
    help is when you feel the need of the social stimulation of a class
    or a group.

6   Learn beyond the point necessary for immediate recall. We all learn
    many things we do not really wish to remember—and which we
    promptly forget. . . . If you want to remember something perma-
    nently, however, you must do what the psychologists call *over-learn-
    ing*. Even after you can recall the fact or perform the skill perfectly,
    you should keep on reviewing it.

7   Use psychological as well as logical practices. You have already had
    an illustration of this rule. In Chapter 1 you were urged to first
    skim this book, then to read it, and then to examine it closely. Now

it seems illogical to many people not to go through a book thoroughly, digesting a paragraph at a time. Yet research has shown that the way recommended here is better.

LEARN FOR LIFE ENRICHMENT

Career-related learning is the key to *career-success;* life enrichment learning is the key to *life fulfillment.* Without both aspects in your learning program, you end up like a hermit crab—overdeveloped in one claw and underdeveloped in the other claw, awkward and out of balance. This example illustrates the value of life enrichment learning:

> "During my engineering studies there was no time for that 'upper-campus' stuff—literature, philosophy, religion, and the like. I smugly shrugged such things off as unimportant—how does understanding Chaucer help you on the job? Since then, I have derived my greatest pleasure from 'back-filling' through a reading program which exposes me to what I missed in college. The ideas of great minds have helped shape my own philosophy. I've gained an understanding of how my profession fits into the broader social, cultural, and historical context. My schooling and work gave me great *training*, but I acquired my education by my own efforts."

This comment sums it up very well. A life enrichment learning program develops the whole person in a way that job-related training and education simply cannot do.

Your local resources for life-enrichment learning are many. Community activities, local college programs, cultural events, reading, hobbies are just a few ways to enrich the quality of your life. Be creative in fashioning learning strategies involving friends and family.

Local clubs, volunteer associations, civic groups, church organizations, and similar groups are excellent vehicles for enrichment learning. Such organizations provide worthwhile community services, and are ideal for developing leadership, management, planning, budgeting,and other skills useful on the job.

The Magic of Reading

In these days of electronic media, reading for pleasure and growth is becoming a lost art. Active reading is a handy, inexpensive, and powerful tool with unlimited perspectives for learning.

Sixteenth-century English philosopher Sir Francis Bacon observed that some books are to be tested, others are to be digested, and a few to be chewed. Your first task is learning which is which, and to help you do that, start your reading program with *How to Read a Book,* referenced at the end of this chapter. The strategies in that book will vastly increase your reading productivity.

Figure 15–2 shows a reading program format I have used for years. My objective is to read an average of one book weekly. I keep a running list of possible books to read, and try to maintain balanced reading in four topic areas. Monitoring progress by using a trend line makes a "game" out of it and keeps me on track.

DEVELOP YOUR LEARNING PLAN

Earlier I suggested four categories for career and life learning. The first step in developing your program is to jot down a list of topics of interest in each category. Your list needn't be more detailed than the example shown in Figure 15–3. The next steps are to:

1   Identify, as specifically as possible, what you wish to learn.
2   Define and "feel" the benefits from this learning. Visualizing the payoffs increases your motivation to follow through.
3   Clarify the conditions which will exist when you reach your learning objective. Identify how you will know when you have acquired sufficient skills or knowledge.
4   Consider alternative learning approaches. Select specific activities and estimate the resources required. Set up interim targets and checkpoints.
5   "Contract" with yourself to invest the resources required.

A learning contract is a specific plan and agreement with yourself to invest the necessary effort, time, and resources to achieve your learning objectives.

The format in Figure 15–4 suggests one way to plan your learning. Use this format to clarify your learning objectives and consider alternative techniques. Estimate how much time and money it will take; "contract" with yourself to make this resource investment. Share these plans with a spouse or mentor. They can give you ideas and support, and help check your progress.

The intriguing thing about lifetime learning is that no one can force you to do it. It's a matter of personal choice. There are enough other activities to fill up your time that it will not happen unless you decide to assign continual learning a priority in your life. Sir John Powell once observed, "He who has no inclination to learn more will be very apt to think that he knows enough." And that can be very hazardous to your career health and life satisfaction.

Figure 15–2   Reading program plan

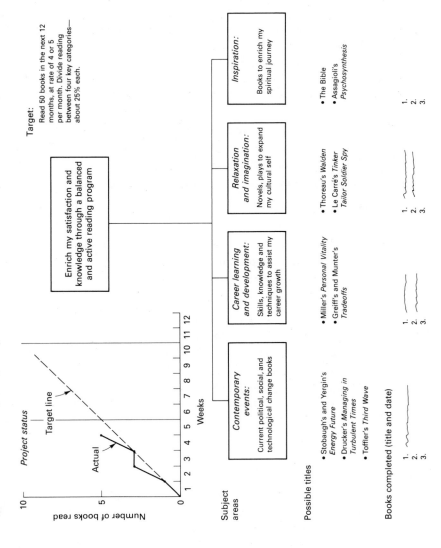

Figure 15–3 A sample set of learning objectives

A   **Professional development in current job/field**

Learn to apply PERT/CPM techniques for scheduling technical programs.

Learn practical techniques for motivating individuals over whom I have no direct control.

Learn project organizing techniques (task force and matrix management methods).

Upgrade my knowledge of software quality assurance and configuration management.

B   **Professional development in future areas of interest**

Master the basics of accounting (need if start my own business).

Learn about office of the future implications.

Learn about accessing and utilizing public data bases.

C   **Personal development**

Organize my time better.

Listen carefully to other people even when I think they're wrong.

Improve my ability to work under pressure.

Learn to accept criticism of my ideas without getting defensive.

D   **Interest expansion**

Learn more about how the U.S. economic system works, and stock/bond/commodities investment alternatives.

Learn modern queen-pawn chess openings.

Learn to coach a youth soccer team to a championship season.

Learn more about technology development during the Renaissance.

SUMMARY OF KEY POINTS

- Lifelong learning is vital for career and life success. Ignoring continuing learning invites career obsolescence and diminishes personal vitality.

- Use a variety of formal, on-the-job, and self-directed learning techniques. Experiment to discover methods which work best for you.

- Analyze your career-related learning needs by examining how technological change and company strategy affect your job, now and in the future.

Figure 15–4 Learning contract

My learning objective: _____

_____

Achievement measures: My learning objective will be achieved when _____

_____

_____

Alternative learning methods

My ideas   *Other peoples' ideas*
Self-development:

_____   _____

_____   _____

_____   _____

On the job:

_____   _____

_____   _____

_____   _____

Formal training:

_____   _____

_____   _____

Action plan

Activity	Target completion date	Follow-up review
_____	_____	_____
_____	_____	_____
_____	_____	_____
_____	_____	_____

Agreement with myself

I plan to complete this learning by _____ (date) and am willing to invest
up to _____ hours and _____ dollars as necessary.

_____   _____
(My signature, indicating commitment)   (Date)

- Life-enrichment learning pays big dividends. Begin by defining all the things you are curious about or have always wanted to learn.
- Develop learning contracts with yourself and/or others. Assign your continual learning program the same priority as tasks you *must* accomplish.

QUESTIONS TO CONSIDER

1   How rapidly is your job/field/organization changing? How much time and effort is needed to keep current? What are the best information sources and methods for you to stay current?

2   How much new learning does your job provide? How could you restructure it to provide more?

3   How much of your own time and money do you currently invest in your continuing education? How much are you willing to invest?

4   What have been your most significant learning experiences during the past year? What techniques do successful colleagues in your organization use to keep up-to-date?

5   What portion of your learning involves life enrichment? What are some new interests you would like to explore? What is the best way to go about it?

RECOMMENDED READINGS

Adler, Mortimer J. and VanDoren, Charles. *How to Read a Book.* New York: A Touchstone Book, Simon & Schuster, Inc., 1972. First published in 1940, this "must-read" book shows how to get maximum value from your reading. Describes how to analyze a book and rapidly extract the main ideas. Specific techniques for reading science and mathematics, practical books, literature, and more.

Knowles, Malcolm. *Self-Directed Learning: A Guide for Learners and Teachers.* Chicago: Follett Publishing Co., 1975. A short and stimulating guide to your learning efforts, based on principles of adult learning. Good discussion of learning contracts.

Kaufman, H. G., ed. *Career Management: A Guide to Combating Obsolescence.* New York: John Wiley & Sons, Inc., 1975. Thirty-seven different articles and papers discussing the nature of technical obsolescence. Helpful guidelines for career management written for the experienced scientist or engineer.

CHAPTER 16

Putting It All Together

Whatever you can do, or dream you can, begin it.
Boldness has genius, power, and magic in it.

Goethe

OVERVIEW

This chapter summarizes the book's key career planning and management concepts. Use it to crystallize your own thinking. Review these concepts again in the future, especially at career/life decision points. It discusses:

- Ten important points for building your strategy
- What an effective strategy should contain
- Ways to test and strengthen your strategy
- Tips for converting your plans into action

BUILD FROM THESE CONCEPTS

Here are the ten most important strategic concepts drawn from the previous fifteen chapters.

1 *Build from your strengths* Concentrate on your strengths. Don't try to be an expert in everything nor attempt to eliminate all your weaknesses. Understand what you do uniquely well and devise opportunities to use and expand these capabilities.

2 *Develop a distinctive mix of transferable skills* Your transferable

skills, knowledge, and capabilities give you the ability to adapt and effectively perform a variety of jobs. Package your skills and knowledge to *differentiate* yourself, and stand out in some important sense.

3 *Give your best at all times* Create a reputation for excellence in all that you do. Work smart, not hard. Think of Pareto's Law—concentrate on the 20% which provides 80% of the value. By doing so, you develop a track record as a superior performer and multiply your future options.

4 *Think "we," not "me"* Create productivity payoffs and benefits for your company, community, and country. If we all work to expand the size of the pie, everyone's share will be bigger. Increase organization efficiency and productivity where you are. An old Dutch proverb says it nicely: bloom where you are planted.

5 *Keep an eye on the future* Anticipate probable changes and prepare for the risks and opportunities which change creates. Monitor technical, social, economic, and political trends which affect your career and life. Identify and take charge of those factors you can control.

6 *Let your success definitions evolve* Experience satisfaction and success throughout your life. Replace obsolete success definitions with new, motivating ones geared to your changing values and desires. Each life stage presents unique opportunities to learn/love/experience/grow/achieve/enjoy.

7 *Set—and reset—stretching goals and objectives* Set goals and objectives throughout life. Make them ambitious; let them call for your best. Maintain goal balance: include personal/family/spiritual/community goals as well as work goals. Keep an ongoing inventory of challenging, bite-size projects. Periodically review, revise, and update your program. Replace obsolete goals with new, motivating dreams.

8 *Learn throughout your life* Acquire new knowledge for current and future career success. Learn for life enrichment as well as career progress. Continue to learn and grow by absorbing new values, skills, knowledge, and attitudes. The world is full of possibilities awaiting your discovery.

9 *Build your interpersonal skills* Use every human encounter as an opportunity to apply good "people skills." Understand your impact on others; be sensitive to unspoken feedback. Learn to enrich others' self-esteem, and you will be respected and liked.

10 *Trust your feelings* Listen to and trust what you feel as well as what you think. View the world as your laboratory for personal growth and discovery. Test cause and effect, figure out the "inner laws" affecting the behavior of your company, colleagues, and self. Respond to your gut and to your head; if they conflict, trust your gut.

TIE YOUR STRATEGY TOGETHER

For 15 chapters I have stressed the importance of developing a career strategy without being terribly specific about what a strategy should contain. The ambiguity was deliberate. Career strategy is not a set of forms to fill out; it is an active thinking process for contemplating possibilities, making decisions, and taking actions. I didn't want to constrain your own thinking process.

But the time has come for pinning down what a good strategy should contain. At a minimum, your strategy should include the following six components. You can develop them in different ways and may want to add others.

- *Personal satisfaction criteria* Develop, review, and periodically update your "must-have" and "nice-to-have" goals, preferences, and values. Use these to improve your decision making and define creative ways to experience success.

- *Trend monitoring* Track key trends in your industry/company/ profession/specialty which affect your career future. Keep abreast of appropriate new information sources and data elements. Pay special attention to trends which create risks or opportunities and respond accordingly.

- *Flexible future job/career objectives* Define future objectives in "functional" terms. Stay flexible; don't focus too narrowly. Specify the nature of the work, key skills, and the organization/function preferences. Narrow them down to short-term targets when you prepare to change jobs. Use your present job to prepare for the future.

- *Current work effectiveness plan* Give and get the most from your work. Know what your job requires, negotiate changes which will benefit you and your employer. Build positive, productive climates. Behave as if you owned the company. Plan your work, work your plan.

- *Learning and growth plan* Acquire new skills and knowledge for career progress and personal enrichment using methods that work best for you. Develop learning contracts; invest the time, energy, and financial resources needed, and give your learning program high priority.

- *Progress monitoring* Periodically review where you've been, where you are now, where you want to go. Maintain a work log; update your accomplishments on a weekly or monthly basis. Thoroughly review your progress at least annually to define new directions, modify your strategy, set new goals. Make *tactical* adjustments to your plan on a daily basis; make *strategic* adjustments following thorough reviews.

TEST YOUR STRATEGY

Strategy development is a continuous, evolutionary process, with periodic adjustments as you and the world change. The purpose of strategy is to provide a framework for developing action plans. Test your strategy against the following criteria to identify ways it might be improved.

External Environment Test

- Is my strategy sound in light of current and probable future trends?
- What major environmental assumptions are implicit in my strategy? Are these valid assumptions?
- Have I identified ways to monitor trends affecting my career? Have I defined specific data and data sources?
- What "leading indicators" of major risk and opportunity is my strategy especially sensitive to?

Organization/Job Test

- Does my organization have a coherent strategy for the future?
- How do my skills/knowledge/interests match my organization's future needs?
- How is my current job preparing me for future opportunities? How can I change it to do more?
- Am I comfortable with my organization's personality?
- Does my job enable me to develop transferable skills?
- Is my job able to grow and change as I grow?

Mobility/Flexibility Test

- How well prepared am I for cross-field mobility?
- What other work functions could I switch to if I wanted to or had to?
- Have I clarified my geographic preferences and constraints?
- Does my strategy increase the number of future job/career options?
- Does it provide sufficient flexibility to shift as external conditions change?
- Have I increased career risk by overspecialization?

Personal/Family Fit Test

- Are my career goals consistent with my family and personal needs?
- Do they adequately balance the different facets of my life? Do they make excessive demands on my family and private life?
- Is my strategy consistent with key value, goals, motivations?

Learning/Developing Test

- Does my strategy include areas for skill/knowledge building?
- Does my strategy build skills/knowledge of high future demand?
- Have I established learning contracts with myself?
- Am I interested in learning what I must learn to stay successful in my work?
- Does my learning program include areas of personal interest and life enrichment?

Practicality/Flexibility Test

- Can my strategy be broken down into specific objectives and action plans?
- Is my strategy capable of being tactically implemented?
- Is it realistic and flexible?
- Does my strategy involve periodic decision making, and permit adjustments based on experience?
- Am I sufficiently motivated to take the necessary actions to make it succeed?

PUTTING YOUR PLANS INTO PRACTICE

In the preface I made three assumptions about you: you are intelligent and curious, your career is important, and you're willing to use strategic thinking to develop career plans and actions. If these initial assumptions about you were true, and you have persevered through all 16 chapters, my guess is that you:

Learned something about yourself, your needs, values, and goals

Answered some questions about yourself, and perhaps raised some additional questions

Identified some current and future career objectives

Gained a better understanding of how your current organization "fits" you

Identified ways to enrich your satisfaction, on and off the job

Started a thinking process to use throughout your life to guide your growth and progress

Generated some new ideas you want to try immediately

The best strategy in the world is useless unless it yields to specific objectives and practical plans you can put into practice, this month, this week, today.

Begin with small steps. Don't try to make it happen all at once. Use a gradual approach—put it into practice a bit at a time. Be content with small victories at first, and create a cumulative process of growth and development.

The starting point for all future success is your current work and life situation. Experiencing lifelong challenge, growth, and satisfaction is entirely within your control. In the final analysis, the quality of your life will not be determined by the infrequent big victories we all relish, but by the small victories, the many things you can do today, and everyday, which add up to an enriching, vital, successful life.

APPENDIX I

SCIENTIFIC/ENGINEERING EMPLOYMENT STATISTICS

This appendix presents data compiled by the National Science Foundation and breaks down scientific/engineering employment by field, sex, employing sector, and primary work activity. This material is from *Science and Engineering Personnel: A National Overview*, published every two years by the National Science Foundation. Single copies are available free from the National Science Foundation, Washington, D.C., 20550.

Employment by field, sex, and primary work activity

Field	Total	Research and development			Management			Teaching	Consulting	Production/ inspection	Reporting, statis. work, computing	Other activities	No report
		Basic research	Applied research	Development	Total	Of R&D	Other R&D						
Total, all fields	685,300	132,500	145,600	407,300	623,000	228,200	394,800	225,200	122,800	353,200	307,000	118,600	38,400
Men	624,100	104,900	125,800	393,500	596,000	218,400	377,700	179,900	113,800	338,400	247,500	107,500	34,500
Women	61,100	27,500	19,800	13,800	26,900	9,800	17,100	45,200	8,900	14,900	59,500	11,100	3,900
Physical scientists	94,500	32,500	33,900	28,000	45,600	28,600	16,900	25,800	3,900	27,600	7,900	3,500	3,600
Men	86,100	28,300	31,400	26,400	44,300	28,000	16,300	24,400	3,300	25,500	7,100	3,200	3,500
Women	8,400	4,300	2,500	1,600	1,300	600	600	1,400	600	2,100	800	300	100
Mathematical scientists	16,300	7,300	5,400	3,600	15,400	6,800	8,600	29,300	1,800	2,600	20,900	1,100	1,200
Men	14,000	7,200	3,200	3,600	14,600	6,500	8,100	25,600	1,700	2,600	10,300	1,100	1,100
Women	2,300	100	2,200	(1)	800	300	500	3,700	(1)	(1)	10,600	(1)	200
Computer specialists	33,900	1,000	4,700	28,200	34,300	14,300	20,000	6,700	11,500	9,200	128,400	5,900	4,000
Men	29,000	1,000	4,300	23,700	31,700	13,200	18,500	5,600	9,800	8,500	101,700	4,700	2,400
Women	5,000	100	500	4,400	2,600	1,100	1,500	1,100	1,700	700	26,700	1,200	1,600
Environmental scientists	26,000	7,500	13,100	5,500	11,600	4,500	7,100	6,300	3,800	8,400	10,700	4,200	1,300
Men	23,100	6,800	10,900	5,300	11,200	4,200	7,100	5,900	3,800	7,000	8,400	4,100	1,100
Women	3,000	700	2,100	200	400	300	(1)	400	(1)	1,400	2,200	100	200
Engineers	378,100	8,500	41,800	327,800	372,600	125,200	247,400	25,100	67,500	257,300	94,400	52,000	21,500
Men	372,000	8,200	40,100	323,700	370,600	123,800	246,800	25,000	67,000	252,000	89,700	51,000	21,400
Women	6,200	300	1,700	4,200	2,000	1,300	600	100	500	5,000	4,700	1,000	100
Life scientists	98,600	59,500	29,900	9,300	69,800	22,500	47,300	56,100	7,700	33,400	9,100	13,000	3,300
Men	70,700	40,100	23,800	6,800	61,600	19,300	42,300	37,500	6,800	30,700	6,800	11,000	2,600
Women	27,900	19,300	6,000	2,500	8,300	3,200	5,100	18,500	900	2,600	2,400	2,000	600
Psychologists	11,900	4,000	7,400	500	20,500	7,800	12,600	29,100	18,100	6,000	10,200	24,100	1,000
Men	8,500	2,500	5,700	300	16,000	6,200	9,800	17,600	14,300	4,400	6,800	21,300	800
Women	3,400	1,500	1,700	200	4,400	1,600	2,800	11,500	3,900	1,500	3,400	2,800	200
Social scientists	26,000	12,100	9,400	4,400	53,300	18,500	34,900	46,900	8,500	8,800	25,400	14,800	2,500
Men	20,900	10,800	6,400	3,700	46,000	17,200	28,800	38,300	7,100	7,800	16,700	11,100	1,600
Women	5,100	1,400	3,000	700	7,300	1,300	6,100	8,600	1,300	1,000	8,600	3,600	900

[1] Too few cases to estimate.

Note: Detail may not add to total because of rounding.

Source: National Science Foundation

Employment by field, sex, and type of employer

Field	Total	Business/ industry	Educational institutions	Nonprofit orgs.	Federal Gov't.	Military	State and local gov't.	Other gov't.	Other	No report
Total, all fields	2,473,200	1,528,100	380,800	80,000	205,800	20,600	145,300	58,300	16,600	37,600
Men	2,214,700	1,445,300	304,800	60,500	187,300	20,300	122,400	55,500	10,800	35,100
Women	231,500	82,700	76,000	19,600	18,600	200	23,000	2,800	5,900	2,500
Physical scientists	212,400	116,300	55,500	7,900	18,000	700	5,200	4,400	700	3,700
Men	197,400	108,400	51,500	7,000	16,900	700	4,600	4,100	600	3,500
Women	15,000	7,900	4,000	900	1,100	(1)	600	300	100	200
Mathematical scientists	88,400	34,200	35,100	3,100	9,400	800	3,300	1,500	(1)	1,000
Men	70,900	25,600	28,600	2,600	8,800	700	2,300	1,300	(1)	900
Women	17,500	8,600	6,500	500	600	(1)	1,000	200	(1)	100
Computer specialists	233,900	173,000	17,900	11,100	14,600	2,900	6,800	3,700	1,100	3,200
Men	193,400	145,100	13,900	9,000	12,300	2,700	4,100	3,500	800	2,200
Women	40,600	27,800	4,000	2,000	2,300	200	2,700	300	300	1,000
Environmental scientists	72,300	40,400	12,900	1,100	10,400	100	4,900	1,800	100	600
Men	64,600	36,000	11,300	1,000	9,500	100	4,400	1,700	100	600
Women	7,700	4,400	1,600	100	900	(1)	500	100	(1)	(1)
Engineers	1,268,400	985,400	48,700	17,900	90,600	11,600	52,900	34,700	3,000	23,600
Men	1,248,500	969,100	47,700	17,800	89,200	11,600	52,600	34,300	3,000	23,200
Women	19,800	16,300	900	100	1,400	(1)	300	400	(1)	400
Life scientists	291,000	86,400	94,400	18,500	41,800	1,800	31,400	4,200	8,700	3,600
Men	227,800	77,300	65,500	9,700	35,500	1,800	27,700	3,500	3,900	3,000
Women	63,200	9,100	28,900	8,900	6,400	(1)	3,800	700	4,800	600
Psychologists	120,900	31,600	55,300	10,200	4,000	2,200	13,200	1,700	2,000	700
Men	89,700	28,500	36,000	7,800	3,100	2,200	8,700	1,400	1,400	600
Women	31,200	3,000	19,400	2,400	900	(1)	4,500	300	600	100
Social scientists	186,000	60,800	61,300	10,200	17,000	500	27,600	6,200	1,000	1,200
Men	149,500	55,300	50,200	5,600	12,000	500	18,000	5,700	1,000	1,100
Women	36,500	5,600	11,100	4,600	5,000	(1)	9,600	500	100	100

[1]Too few cases to estimate.
Note: Detail may not add to total because of rounding.
Source: National Science Foundation.

APPENDIX II

LONG-RANGE TECHNICAL EMPLOYMENT FORECASTS

The information in this appendix, compiled by the U.S. Bureau of Labor Statistics, forecasts technical employment through 1990 in 43 technical specialties. These projections are based on anticipated federal R & D budgets, defense expenditures, industrial expansion needs, and demographic changes.

A warning about projections: don't build your strategy around numbers alone. The data are no better than the assumptions made at the time the data were compiled. Changes in technology, international conditions, government policy, economic factors, and dozens of unpredictable events can markedly affect data validity.

Changing Employment Between 1978 and 1990

If the statement reads . . .	Employment is projected to . . .
Much faster than average growth	Increase more than 50 percent
Faster than average growth	Increase 25 to 49.9 percent
Growth about as fast as average	Increase 15 to 24.9 percent
Growing more slowly than average	Increase 5 to 14.9 percent
Little change	Increase or decrease no more than 4.9 percent
Decline	Decrease 5 percent or more

Opportunities and Competition for Jobs

If the statement reads . . .	The demand for workers may be . . .
Excellent opportunities	Much greater than the supply
Very good opportunities	Greater than the supply
Good or favorable opportunities	About the same as the supply
May face competition	Less than the supply
Keen competition	Much less than the supply

Occupation	Estimated employment, 1978	Average annual openings, 1978-90[1]	Employment prospects
Computer and related occupations			
Computer operating personnel	666,000	12,500	Employment of console and peripheral equipment operators expected to rise about as fast as average as use of computers expands. Employment of keypunch operators expected to decline, however, due to more efficient direct data entry techniques.
Programmers	247,000	9,200	Employment expected to grow faster than average as computer use expands, particularly in accounting, business management, and research and development. Brightest prospects for college graduates with degree in computer science or related field.
Systems analysts	182,000	7,900	Employment expected to grow faster than average as computer capabilities are increased and computers are used to solve a greater variety of problems. Excellent prospects for graduates of computer-related curriculums.
Conservation occupations			
Foresters	31,200	1,400	Employment expected to grow about as fast as average as environmental concern and need for forest products increase. But, applicants likely to face keen competition for jobs.
Forestry technicians	13,700	700	Employment expected to grow faster than average as technicians increasingly do tasks formerly handled by foresters. Even applicants with specialized postsecondary school training may face competition, however.
Range managers	3,700	200	Employment expected to grow faster than average. Good employment prospects likely as use of rangelands for grazing, recreation, and wildlife habitats increases.

Occupation	Estimated employment, 1978	Average annual openings, 1978-90[1]	Employment prospects
Soil conservationists	9,300	450	Employment expected to increase as fast as average as organizations try to preserve farmland and comply with recent conservation and antipollution laws. Competition may be keen.
Engineers			
Engineers	1,136,000[2]	46,500[2]	Employment expected to grow slightly faster than average. Good employment opportunities for graduates with an engineering degree.
Aerospace engineers	60,000	1,900	Employment expected to grow about as fast as average due to limited increase in Federal expenditures on space and defense programs.
Agricultural engineers	14,000	600	Employment expected to grow faster than average in response to increasing demand for agricultural products, modernization of farm operations, and increasing emphasis on conservation of resources.
Biomedical engineers	4,000	175	Employment expected to grow faster than average, but actual number of openings will be small. Increased research funds could create new jobs in instrumentation and systems for delivery of health services.
Ceramic engineers	14,000	550	Employment expected to grow faster than average as a result of need to develop and improve ceramic materials for nuclear energy, electronics, defense, and medical science.
Chemical engineers	53,000	1,800	Employment expected to grow about as fast as average. Growing complexity and automation of chemical processes will require additional chemical engineers to design, build, and maintain plants and equipment.

Occupation	Estimated employment, 1978	Average annual openings, 1978-90[1]	Employment prospects
Civil engineers	155,000	7,800	Employment expected to increase about as fast as average as result of growing need for housing, industrial buildings, electric power generating plants, and transportation systems. Work related to environmental pollution and energy development will also cause growth.
Electrical engineers	300,000	10,500	Employment expected to increase about as fast as average due to growing demand for computers, communications equipment, military electronics, and electrical and electronic consumer goods. Increased research and development in power generation also should create openings.
Industrial engineers	185,000	8,000	Employment expected to grow faster than average due to industry growth, increasing complexity of industrial operations, expansion of automated processes, and greater emphasis on scientific management and safety engineering.
Mechanical engineers	195,000	7,500	Employment expected to increase about as fast as average due to growing demand for industrial machinery. Need to develop new energy systems and to solve environmental pollution problems will also cause growth.
Metallurgical engineers	16,500	750	Employment expected to grow faster than average due to need to develop new metals and alloys, adapt current ones to new needs, solve problems associated with efficient use of nuclear energy, and develop new ways of recycling solid waste materials.
Mining engineers	6,000	600	Employment expected to grow much faster than average due to efforts to attain energy self-sufficiency and to develop better mining systems.

Occupation	Estimated employment, 1978	Average annual openings, 1978-90[1]	Employment prospects
Petroleum engineers	17,000	900	Employment expected to grow faster than average as demand for petroleum and natural gas requires increased drilling and more sophisticated recovery methods.
Environmental scientists			
Geologists	31,000	1,700	Employment expected to grow faster than average as domestic mineral exploration increases. Good opportunities for persons with degrees in geology or earth science.
Geophysicists	11,000	600	Employment expected to grow faster than average as petroleum and mining companies need additional geophysicists who are able to use sophisticated electronic techniques in exploration activities. Very good opportunities for graduates in geophysics or related areas.
Meteorologists	7,300	300	Employment expected to increase about as fast as average. Favorable opportunities for persons with advanced degrees in meteorology. Others expected to face competition.
Oceanog-raphers	3,600	150	Although employment expected to grow about as fast as average, competition for openings is likely. Best opportunities for persons who have a Ph.D.; those who have less education may be limited to research assistant and technician jobs.
Life science occupations			
Biochemists	20,000	900	Employment expected to grow faster than average due to increase in funds for biochemical research and development. Favorable opportunities for advanced degree holders.
Life scientists	215,000	11,200	Employment expected to grow faster than average due to increasing

Occupation	Estimated employment, 1978	Average annual openings, 1978-90[1]	Employment prospects
			expenditures for medical research and environmental protection. Good opportunities for persons with advanced degrees.
Soil scientists	3,500	180	Little employment growth expected. Applicants may face competition for jobs.
Mathematics occupations			
Mathematicians	33,500	1,000	Slower than average employment growth is expected to lead to keen competition for jobs. Opportunities expected to be best for advanced degree holders in applied mathematics seeking jobs in government and private industry.
Statisticians	23,000	1,500	Employment expected to grow faster than average as use of statistics expands into new areas. Persons combining knowledge of statistics with a field of application, such as economics, may expect favorable job opportunities.
Physical scientists			
Astronomers	2,000	40	Little change in employment is expected as only slight increases in funds for basic research in astronomy are expected. Competition for jobs is likely to be keen.
Chemists	143,000	6,100	Employment expected to grow about as fast as average as a result of increasing demand for new products and rising concern about energy shortages, pollution control, and health care. Good opportunities should exist, except for academic positions.
Physicists	44,000	1,000	Although employment will grow more slowly than average, generally favorable job opportunities are expected for persons with advanced

Occupation	Estimated employment, 1978	Average annual openings, 1978-90[1]	Employment prospects
			degrees in physics. However, persons seeking college and university positions, as well as graduates with only a bachelor's degree, will face keen competition.
Social Scientists			
Anthropologist	7,000	350	Employment expected to increase about as fast as average. Nearly all new jobs will be in nonacademic areas. Even persons with a Ph.D. in anthropology can expect keen competition for jobs.
Economists	130,000	7,800	Employment expected to grow faster than average. Master's and Ph.D. degree holders may face keen competition for academic positions but can expect good opportunities in nonacademic areas, particularly for those trained in quantitative methods. Persons with bachelor's degrees likely to face keen competition.
Geographers	10,000	500	Employment expected to grow about as fast as average. Advanced degree holders likely to face keen competition for academic positions, but good prospects in nonacademic areas. Bachelor's degree holders will face competition.
Historians	23,000	700	Little change in employment expected. Keen competition anticipated, particularly for academic positions. Best opportunities for Ph.D.'s with a strong background in quantitative research methods.
Political scientists	14,000	500	Employment expected to increase more slowly than average. Keen competition likely, especially for academic positions. Best opportunities for advanced degree holders with training in applied fields such as public administration or public policy.

Long-Range Technical Employment Forecasts   **205**

Occupation	Estimated employment, 1978	Average annual openings, 1978-90[1]	Employment prospects
Psychologists	130,000	6,700	Employment expected to grow faster than average. Graduates face increasing competition, particularly for academic positions. Best prospects for doctoral degree holders trained in applied areas, such as clinical, counseling, and industrial psychology.
Sociologists	19,000	600	Employment expected to grow more slowly than average. Ph.D.'s may face competition, particularly for academic positions. Best opportunities for Ph.D.'s trained in quantitative research techniques. Very keen competition below Ph.D. level.

Other scientific and technical occupations

Occupation	Estimated employment, 1978	Average annual openings, 1978-90[1]	Employment prospects
Broadcast technicians	40,000	[3]	Employment expected to increase about as fast as average as new radio and television stations are licensed and as cable television stations broadcast more of their own programs. Job competition is keen, however, and prospects are best in small cities.
Drafters	296,000	11,000	Employment expected to grow about as fast as average due to increasing complexity of designs of modern products and processes. Best prospects for graduates with associate degrees in drafting.
Engineering and science technicians	608,000	23,000	Employment expected to grow faster than average as more technicians will be needed to assist the growing number of engineers and scientists. Favorable job opportunities, particularly for graduates of postsecondary school training programs.
Food technologists	15,000	500	Employment expected to grow about as fast as average due to increasing demand for food technologists in

Occupation	Estimated employment, 1978	Average annual openings, 1978-90[1]	Employment prospects
			research and development, quality control, and production. Favorable opportunities for persons with food technology degrees.
Surveyors and surveying technicians	62,000	2,300	Employment expected to grow about as fast as average due to increased construction activity. Job opportunities are affected by economic conditions.

[1] Due to growth and replacement needs. Does not include transfers out of occupations. Estimates of replacement openings on working life tables developed by Bureau of Labor Statistics.
[2] Total is not sum of individual estimates because all branches of engineering are not covered separately in the *Occupational Outlook Handbook*.
[3] Estimate not available.

INDUSTRY MEDIANS

	Yardsticks of management performance											
	Profitability								Growth			
	Return on equity			debt/equity ratio	Return on total capital			net profit margin	Sales		Earnings per share	
Industry	5-year average	5-year rank	latest 12 months		latest 12 months	5-year rank	5-year average		5-year average	5-year rank	5-year average	5-year rank
Broadcasting	22.4%	1	16.1%	0.3	13.4%	1	17.9%	6.0%	14.3%	21	23.7%	1
Drugs	21.3	2	16.9	0.1	13.6	8	15.5	8.5	12.2	31	11.0	30
Tobacco	20.6	3	23.6	0.5	14.1	13	13.6	6.3	15.4	15	16.6	10
Office equipment and services	20.3	4	17.5	0.3	13.4	7	15.7	3.9	18.0	8	17.8	7
Office equipment	18.8	–	12.1	0.3	9.5	–	14.7	3.5	15.4	–	11.0	–
Office services	25.1	–	19.1	0.2	15.1	–	18.9	3.9	24.3	–	24.8	–
Insurance	19.8	5	14.7	0.1	13.3	4	16.4	7.4	14.4	20	18.9	5
Life and accident	13.8	–	13.2	0.1	11.9	–	13.4	9.7	11.3	–	12.4	–
Fire and casualty	22.8	–	17.2	0.2	15.6	–	21.7	7.5	16.0	–	26.7	–
Multiple line	20.0	–	14.9	0.1	13.5	–	16.8	6.5	14.1	–	22.0	–
Publishing	19.8	5	18.0	0.2	13.9	2	17.2	6.2	15.7	14	16.8	9
Construction, mining, transport equip.	19.7	7	17.1	0.4	14.4	12	13.7	6.5	14.2	23	15.6	15
Construction and materials handling	13.0	–	10.4	0.4	8.6	–	9.9	4.1	12.6	–	5.6	–
Mining and drilling equipment	21.7	–	25.1	0.3	19.9	–	15.4	10.5	27.1	–	24.4	–
Rail equipment	12.6	–	12.9	0.7	5.6	–	7.1	3.3	12.4	–	6.9	–
Energy	19.5	8	18.3	0.4	11.5	15	12.8	4.4	21.0	3	16.5	11
International oils	19.4	–	19.9	0.2	16.1	–	14.6	4.4	17.0	–	16.4	–
Other oil and gas	20.2	–	18.1	0.5	11.2	–	12.9	4.4	22.9	–	16.7	–
Coal	9.7	–	4.7	0.4	3.8	–	6.6	1.2	10.7	–	–6.3	–
Toiletries and cosmetics	19.5	8	15.4	0.2	11.9	9	15.1	7.1	12.9	30	12.0	25
Health care	19.2	10	19.2	0.2	13.1	14	13.4	6.3	20.2	5	21.8	2
Electronics	19.1	11	16.1	0.2	13.2	6	16.3	6.2	17.9	9	21.6	3
Computers	19.0	12	13.7	0.2	11.6	4	16.4	6.3	23.0	2	21.3	4
Aerospace and defense	18.6	13	18.0	0.2	13.0	10	14.4	3.9	15.2	16	18.4	6
Electrical equipment	18.3	14	17.8	0.1	16.2	3	16.5	6.3	15.2	16	15.0	17

Yardsticks of management performance

| | Profitability | | | | | | | | Growth | | | |
| | Return on equity | | | | Return on total capital | | | | Sales | | Earnings per share | |
Industry	5-year average	5-year rank	latest 12 months	debt/equity ratio	latest 12 months	5-year rank	5-year average	net profit margin	5-year average	5-year rank	5-year average	5-year rank
Contractors	18.1%	15	16.9%	0.4	12.9%	18	12.2%	2.5%	17.9%	9	15.2%	16
Heavy construction	16.2	–	16.9	0.3	12.9	–	12.4	1.9	16.9	–	12.6	–
Home-building	17.1	–	9.2	0.5	6.0	–	9.9	2.4	16.4	–	26.7	–
Oilfield drillers	20.7	–	27.9	0.6	17.9	–	12.5	13.6	18.3	–	21.5	–
Leisure	17.7	16	12.6	0.5	10.2	21	11.4	4.6	13.6	25	13.7	20
Hotels and gambling	20.3	–	18.1	1.1	11.0	–	9.9	5.0	13.5	–	20.3	–
Entertainment	20.8	–	20.8	0.3	12.8	–	17.1	7.0	16.4	–	16.3	–
Recreation	12.3	–	9.6	0.3	8.2	–	11.1	4.4	12.9	–	8.2	–
Specialty retailers	17.6	17	17.3	0.6	11.6	15	12.8	2.5	17.3	11	16.9	8
Drug chains	19.1	–	18.6	0.6	12.5	–	14.2	2.4	20.3	–	18.6	–
Fast food chains	15.7	–	12.2	1.0	8.0	–	11.1	1.5	18.1	–	11.7	–
Other specialists	16.1	–	15.6	0.4	10.6	–	13.3	2.7	14.2	–	15.7	–
Industrial equipment and services	17.1	18	15.4	0.3	12.3	11	13.8	5.1	13.7	24	14.9	18
Production equipment	17.1	–	15.1	0.4	10.9	–	14.0	5.1	14.2	–	14.1	–
Specialty equipment and materials	16.8	–	14.6	0.3	12.3	–	13.5	4.9	13.6	–	13.4	–
Industrial services	19.0	–	23.7	0.5	15.9	–	13.7	6.0	17.2	–	22.8	–
Chemicals	16.6	19	14.7	0.4	10.0	23	11.1	5.4	13.5	26	9.9	38
Diversified	13.1	–	13.3	0.4	9.6	–	9.8	4.9	12.6	–	7.9	–
Specialized	18.2	–	15.9	0.4	11.2	–	13.0	6.7	15.0	–	15.3	–
Natural gas	16.5	20	19.1	0.7	9.4	40	8.7	4.7	20.6	4	10.5	34
Producers and pipeliners	18.3	–	21.1	0.7	11.2	–	9.8	5.0	23.0	–	12.7	–
Pipeliners and distributors	18.3	–	22.4	0.5	11.0	–	9.6	5.4	23.8	–	13.1	–
Distributors	13.5	–	14.7	0.8	8.8	–	7.7	4.0	19.6	–	8.4	–
Conglomerates	16.3	21	15.2	0.5	9.3	36	9.9	4.0	11.1	38	13.8	19
Branded foods	15.9	22	15.7	0.4	11.6	19	11.9	3.4	11.9	35	11.8	26
Diversified	16.1	–	15.2	0.4	11.3	–	11.6	3.2	11.3	–	11.9	–
Specialized	15.9	–	17.0	0.3	12.7	–	12.3	3.7	12.1	–	11.6	–
Truckers and shippers	15.5	23	15.4	0.5	9.4	33	10.1	3.9	14.7	18	11.4	27
Truckers	22.2	–	18.9	0.5	8.7	–	10.6	3.7	14.8	–	14.2	–
Other surface transportation	14.7	–	14.5	0.5	11.7	–	10.0	6.7	8.2	–	8.8	–

Yardsticks of management performance

Industry	Profitability — Return on equity				Return on total capital				Growth — Sales		Earnings per share	
	5-year average	5-year rank	latest 12 months	debt/equity ratio	latest 12 months	5-year rank	5-year average	net profit margin	5-year average	5-year rank	5-year average	5-year rank
Wholesalers	15.4%	24	14.7%	0.5	11.2%	20	11.5%	1.2%	14.3%	21	12.2%	24
Food distributors	14.9	–	12.8	0.7	8.7	–	10.6	0.8	14.7	–	10.6	–
Other wholesalers	17.6	–	18.7	0.4	13.2	–	12.8	1.9	14.3	–	15.2	–
Food processors	15.3	25	14.0	0.5	11.2	23	11.1	1.4	8.6	46	11.3	28
Meatpackers	13.9	–	13.3	1.0	10.3	–	10.1	0.9	7.9	–	6.8	–
Agricultural commodities	15.3	–	14.8	0.4	12.3	–	11.4	3.0	11.3	–	12.9	–
Household goods	15.2	26	13.9	0.4	9.9	23	11.1	3.8	10.1	42	10.1	35
Appliances	14.2	–	14.4	0.5	9.1	–	10.8	2.9	8.2	–	8.6	–
Housewares and furnishings	11.3	–	9.7	0.7	7.5	–	9.5	1.9	9.3	–	11.6	–
Housekeeping products	17.4	–	15.6	0.2	11.4	–	13.6	5.4	11.8	–	10.1	–
Supermarkets	15.0	27	14.8	0.9	10.2	32	10.2	1.2	13.1	28	15.7	13
Major chains	18.1	–	17.3	1.0	10.4	–	10.6	1.1	14.0	–	13.8	–
Regional chains	14.4	–	13.2	0.9	9.6	–	10.1	1.2	12.7	–	15.7	–
Financial services	14.9	28	11.9	0.7	8.1	36	9.9	4.0	14.6	19	16.2	12
Beverages	14.7	29	18.9	0.3	11.9	26	10.8	4.9	12.2	31	9.3	42
Forest products	14.7	29	11.9	0.4	8.6	33	10.1	4.7	11.8	36	9.9	38
Building materials	14.6	31	9.8	0.4	7.6	30	10.5	3.8	10.3	41	10.7	32
Lumber	14.5	–	7.8	0.5	6.0	–	9.7	3.6	12.8	–	9.3	–
Cement and gypsum	15.2	–	11.4	0.7	7.5	–	10.7	5.2	9.9	–	14.9	–
Other building materials	12.5	–	9.7	0.4	7.9	–	8.4	2.9	9.6	–	8.8	–
Building equipment	17.0	–	14.0	0.4	12.6	–	14.5	5.4	10.3	–	18.2	–
Multi-industry companies	14.6	31	13.7	0.3	10.0	26	10.8	5.3	9.7	43	12.9	22
Banks	14.4	33	14.1	0.4	11.5	22	11.3	5.6	18.3	7	11.0	30
New York banks	14.4	–	14.1	0.5	12.0	–	11.3	3.8	19.7	–	9.7	–
Regional banks	14.5	–	14.0	0.3	11.3	–	11.7	5.9	17.6	–	11.4	–
Telecommunications	14.3	34	14.5	0.9	7.5	44	7.1	10.3	11.6	37	8.2	43
Auto suppliers	13.9	35	12.4	0.4	9.8	33	10.1	3.3	10.6	39	10.6	33
Tire and rubber	6.9	–	8.6	0.5	6.9	–	6.0	2.4	8.3	–	1.8	–
Parts makers	15.3	–	14.1	0.4	10.1	–	12.2	3.9	13.0	–	13.6	–

Yardsticks of management performance

| | Profitability | | | | | | | | Growth | | | |
| | Return on equity | | | | Return on total capital | | | | Sales | | Earnings per share | |
Industry	5-year average	5-year rank	latest 12 months	debt/ equity ratio	latest 12 months	5-year rank	5-year average	net profit margin	5-year average	5-year rank	5-year average	5-year rank
Brokerage	13.8%	36	17.6%	0.5	15.5%	17	12.7%	4.5%	30.8%	1	11.3%	28
Nonferrous metals	13.8	36	11.6	0.4	9.0	31	10.4	4.0	13.5	26	9.5	41
Autos and trucks	13.4	38	7.3	0.3	6.0	26	10.8	2.2	12.1	34	7.2	44
Railroads	12.7	39	13.2	0.6	8.5	43	8.0	6.3	13.1	28	15.7	13
Packaging	12.5	40	10.1	0.5	8.0	38	9.1	3.0	12.2	31	10.0	36
General retailers	12.1	41	10.6	0.7	7.9	41	8.6	2.2	9.7	43	9.6	40
Department stores	12.0	–	12.1	0.6	8.0	–	8.8	2.6	10.0	–	8.0	–
Discount and variety stores	12.9	–	8.7	0.9	6.9	–	8.5	1.4	9.6	–	11.9	–
Steel	12.1	41	13.9	0.4	10.6	42	8.3	4.4	10.4	40	4.9	45
Electric utilities	11.8	43	12.6	0.9	6.6	45	6.6	10.4	15.8	13	2.7	46
Northeast	11.5	–	12.2	0.9	6.3	–	6.6	9.1	12.9	–	3.3	–
Midwest	11.8	–	11.9	0.9	6.2	–	6.5	11.9	15.1	–	2.1	–
Southeast	12.1	–	13.2	0.9	6.6	–	6.4	10.2	16.1	–	2.9	–
Southwest	14.0	–	14.3	0.9	7.4	–	7.6	12.0	19.8	–	6.0	–
West	11.2	–	13.3	0.9	7.4	–	6.8	10.9	20.3	–	1.2	–
The thrifts	11.8	43	def	0.5	0.9	29	10.6	def	19.8	6	10.0	36
Apparel	11.0	45	11.3	0.3	9.7	38	9.1	3.3	8.8	45	13.2	21
Clothing	12.4	–	12.8	0.4	10.1	–	9.7	3.3	7.7	–	15.5	–
Textiles	9.1	–	9.4	0.3	7.3	–	7.0	2.7	8.7	–	7.9	–
Shoes	23.5	–	23.0	0.3	16.4	–	16.3	4.4	13.3	–	19.0	–
Airlines	9.0	46	5.3	1.7	3.9	46	5.9	1.2	16.5	12	12.7	23
All-industry medians	15.5	–	14.7	0.4	10.9	–	11.1	4.5	14.0	–	12.5	–

EXPLANATION OF YARDSTICKS OF PERFORMANCE

Return on stockholders' equity   Companies obtain their capital from two sources: stockholders and creditors. Return on stockholders' equity is the percentage return on the stockholders' portion of the capital. We express earnings per common share (primary basis) as a percentage of the stockholders' equity per share, assuming conversion of all convertible preferred, at the start of the year. The five-year return in our Yardsticks tables is the average of the returns calculated for the four years 1977 through 1980 and the 12-month period ending with the most recent quarterly report. (For companies whose fiscal years end in January, February, or March, the years 1978–81 are used in place of 1977–80.)

Return on total capital   This figure is the percentage return on a combination of stockholders' equity (both common and preferred) plus capital from long-term debt including current maturities, minority stockholders' equity in consolidated subsidiaries and accumulated deferred taxes and investment tax credits. The profit figure used in this computation is the sum of net income, minority interest in net income and estimated aftertax interest paid on long-term debt—in other words, income before charges (which means, primarily, interest payments on long-term debt) relating to the nonequity portion of the capital.

The return on total capital is a "basic" measure of an enterprise's profitability. For companies that derive all of their capital from common equity, the two profitability measures will be identical. But a company that employs debt wisely can thereby boost its return on stockholders' equity well above the return on total capital. The time periods employed for this calculation are the same as those for return on stockholders' equity.

Debt-to-equity ratio   This ratio tells us to what extent management is using borrowed funds (leverage) in an attempt to increase profits. It is calculated as of the end of the last reported fiscal year by dividing long-term debt (including current maturities) by the sum of stockholders' equity, minority stockholders' interest and accumulated deferred taxes and investment tax credits. A high debt-to-equity ratio makes earnings more volatile and is usually considered prudent only in relatively stable industries.

Net profit margin   This measure gives a view of profits different from either the return on stockholders' equity or the return on total capital. Calculated by dividing net profits for the latest 12 months by net sales, it reveals what percentage of each dollar of revenue is available for payment of dividends and reinvestment in the business.

Earnings-per-share growth   To present a clear picture and even out short-run distortions caused by very poor or very good years, we reach back over ten years to measure five-year earnings-per-share growth.

Using primary earnings per share, we take the average for the most recent five years (1977 through the latest 12 months) and compute the percentage change from the average for the preceding five years (1972 through 1976). We then express that change interms of a five-year compound annual rate of growth, since the midpoints of the two periods are five years apart.

Sales growth   As with earnings growth, we go back ten years to calculate the five-year rate of sales growth. We compare average sales for the company's most recent five years against the average for the preceding five years, and express the change in terms of a five-year compound annual growth rate.

Index